AF862493

"Enjoy"

Col. John Zitzelberger
6719 Beaver Ct.
Midland, GA 31820
706-569-9757

Blood Runs Red, White and Blue

Blood Runs Red, White and Blue

by William H. Abbott

published by
World Liberty Publishing
a division of
World Liberty Foundation, Inc.
479 N. Dixie Avenue
Titusville, Florida 32796

To Bernice

Contents

Foreword

Why another book about the Korean War?

It's time for another fresh look, and this one is not a rehash of previous publications.

Many Americans do not realize that after World War II our occupation of Korea was short-lived, during which time we looked directly across the 38th parallel border at hostile Russian, as well as North Korean, soldiers. Before we withdrew from Korea in 1949 there were many infractions of the peace, and minor hostile actions occurred between their forces and America's.

After four years of occupation the U.S. Army withdrew from Korea and pulled our troops back to Japan, leaving the Korean police force and army to face the heavily armed army of North Korea. William Abbott was part of this occupation and subsequent withdrawal before the war.

The U.S. Army in Japan was ill-equipped, poorly trained, and "Johnsonized" (regiments were reduced by one-third by then Secretary of the Army, Louis Arthur Johnson, famous along with President Harry S Truman for defense budget cuts). William

Abbott was part of these forces, while the North Koreans waited for an opportunity to invade the South and unify the country under their control. This opportunity came in June of 1950, and the Korean War began.

In September of 1950 the U.S. Army's 7th Infantry Division and the 1st Marine Division landed at Inchon, cutting the supply lines of the North Korean army. William Abbott was part of the returning American forces who now had the North Koreans in full flight north to the Yalu River, the border between North Korea and China. The war was all but over. But in November of 1950 the Chinese intervened and a new conflict emerged.

Many Americans and South Koreans were trapped, with long supply lines and no linkage between units. They were vulnerable to annihilation. No place was this more apparent than at the Chosin Reservoir, where a Marine division and an Army regiment were trapped high in the mountains. William Abbott was among them. They had only a narrow corridor of escape toward the sea and, with Chinese intervention, all escape routes were severed and thirty thousand Americans were trapped in a frozen hell. In the Chosin Reservoir there were two enemies: the Chinese, and the freezing cold weather. This battle goes down in history as one of the most epic battles this country ever fought.

The Army east of the Chosin fought until it had nothing left to fight with. When it had expended all its resources, few men survived to fight another day. Many died or were captured to become prisoners of war to the Chinese.

Those who became prisoners endured some of the harshest conditions that American POWs have ever had to endure. We have never, in any of our country's wars, had enemies who treated our prisoners the way the Chinese treated those who were captured. The treatment our men endured in captivity in Korea ranks as some of the most grueling, harsh treatment that our armed forces have endured as POWs, and William Abbott was among them.

Why another book about Korea? This author, William Abbott, lived through all the phases of this war, from the occupation of Korea to the pullback into Japan, through the Korean War, and subsequent incarceration as a POW. He is also one of the few prisoners not just admired by his peers, but also by his captors—they could not subjugate him as they did so many others.

William Abbott is a soldier's soldier. He has thrived on adversity his entire life, and is a testament to patriotism and love of country that few have achieved. His account of this war is well worth study. Indeed, this book is a must-read for any soldier who goes into harm's way, or anyone seeking to be a student or historian of military science and history.

John Zitzelberger
Colonel, U.S. Army, retired

Introduction

This book is a reflection on my life as a U.S. soldier, and my recollection of each event is as vivid in my mind today as though it happened yesterday.

I make no effort to soft-pedal my actions or the actions of others, good or bad. This book is straight from my heart. You, the reader, will have to decide whether I have been a saint, a sinner, or somewhere in between.

As an old sergeant wrote to me after the war, thanking me for helping my fellow POWs, "I'll always remember you as the one person who could be counted on to help others…Just wanted to thank you, Abbott."

This book will show, however, that I have very little patience with those who fail to maintain loyalty to their oath as a soldier. I can think of no circumstance under which a soldier can justify a violation of his oath.

After the war, the Army decorated a few men for their actions during captivity. I was proud to be one of those men.

It is one thing to do the job you are trained to do as a combat soldier—and I make no apologies for my actions there—but it's another thing to maintain allegiance as a POW. For a soldier to remain loyal under oath to the nation for which he has taken that oath when he has an option to take the easy way out—to cooperate with the enemy in order to get along easier—maintaining the oath separates the corn from the husk rather quickly.

Could I have done more? Yes. Up until the battle at the Chosin Reservoir, when things got hot, the war was simply fun and games for me. I could have provided more leadership, but I was not aware of how much the officers and men depended on me until we came under attack and heavy fire.

In many cases the POW was his own worst enemy, as well as causing insurmountable damage to his fellow POWs. I credit the Chinese with their ability to identify their prospective collaborators. The Chinese knew those men who had not been taught to appreciate what it means to be among God's chosen people—an American citizen. Communism was an easy sell for these people, and the Chinese exploited it.

I record in my book a conversation with some soldiers who decided to stay in China after the war ended. One fact was clear to me: the reason they did not return to the United States was fear—fear of being court-martialed. They simply did not have the guts to face a court-martial and assume responsibility for their actions. However, I place the blame for the collaboration of these men, at least in part, on the U.S. government.

Communism had been a force in America since 1921. Some of Roosevelt's and Truman's most trusted aides were Communist agents. Up to, during, and through the Second World War and into Korea, the government knew the hazards posed by the world Communist movement, yet the fact is that the U.S. government supported Communism from its inception.

The U.S. had been able to teach its soldiers how to shoot, how to march around a parade field, and be the best combat soldiers in the world. But no time had been spent on making them into patriots, which is required to function in captivity. It is absolutely essential that they have what I call "that fire in the belly" to uphold their oath as soldiers and as U.S. citizens.

Loyalty is something that cannot be bought with money; it doesn't cost a solitary dime, but the value to mankind is far greater than my ability to convey with pen and paper. Without loyalty, a man has nothing. Without loyalty, nothing else really matters. Without loyalty, a man is like a ship without a rudder, floating around and around in the ocean.

Once when I was in a bar in Washington, D.C. in 1954, a lieutenant from one of the POW camps came in. He told the people at my table that if he had to place his life in the hands of one man, and he had the whole world to choose from, he would choose me. (Others who had been at the Chosin accused me of being too mean to die.)

At Ft. Sill, an old soldier, a World War II ex-POW, approached me and said, "You're Bill Abbott."

"Yes Sir, I am," I said.

"Let me shake hands with God."

I was stunned for a moment. Then he said, "To these men, that's what you are."

I got my composure, and said, "Well, thank you, Sir. You've made my day. If I have contributed in some small way to these brave men, then I know God will bless me for it."

George Pryor, a dear friend who served with me at the Chosin, told my wife recently, in his words, "He done a lot of bad things he should not have done. But we all loved him and we protected him."

Lou Richardson, from Camp 1, told me in a recent conversation, "Abbott, you made us laugh. You made us laugh when there was nothing but death all around us."

I've been known to steal food, and I've been known to steal boiled water and crawl through Chinese lines to help the sick. I've also been known to get drunk on sake on Christmas Eve and preach a sermon on the birth of Christ. But I have never lost sight of loyalty to my country.

Every professional soldier in the world, regardless of country, appreciates loyalty, and the Chinese were no exception.

The Chinese learned early on that I was not going to furnish them propaganda or sign peace appeals. My answer to such things as "Truman was a warmonger" and "MacArthur was a murderer" was very simple: "They might be all of that, Sir. I don't know these gentlemen. One thing I do know—they represent the country to which I am under oath to protect and serve, right or wrong. Now that right or wrong may be subject to debate, but loyalty to my family and fellow Americans is not debatable, Sir."

This book will take you from the Korean occupation in 1948 to Japan, then back to Inchon in South Korea and the landing in September 1950, to the North Korean landing at Iowan, to the Fusan Reservoir, to the Chosin Reservoir, and the battle and the breakout from the Chosin when I became a wounded POW with minimal clothing in minus-sixty-degree weather.

It is about life and action in the POW camps, the "love and grace" of the Chinese, the labor camps, the Korean jail, and the bad side of interrogation; also the reasons behind so many deaths in the camps, surpassing any war in history. But mostly it is a story of one critically wounded soldier whose courage and love of his country, his God, and his fellow man sustained him throughout the dark days of a hell on earth.

My hope for this book is that every person who reads it will come away with a positive belief that each of us has an obligation to maintain allegiance to our God, family and country; and that everyone, regardless of rank or standing can make a difference in his own life, as well as the lives of others.

To the men and officers in uniform, I always say: "Remember that you are the final leg of diplomacy of your nation. After all the political heads have gone home, it is your action upon which the world will judge your country. God has given you an opportunity to be a citizen and a defender of the freest, most prosperous nation on the face of the earth. That freedom has been preserved through the sweat and blood of others. Should you do less for those who follow in your footsteps? I think not. With a lump in your throat, a tear of compassion in your eye and God's hands on your shoulders, be resolute in your heart that freedom shall not perish from this land on your watch. I wish you Godspeed, and may the blood in your veins always run red, white, and blue."

As for me, I hope those of you who know me will say without reservation, "He has worked as hard as any man alive to advance the cause of freedom, and God gave him the insight and the ability to do so. He was always ready to help his fellow man and he swathed a path of truth and justice as he passed through."

I hope one day when I move over into the next world you will wrap my bones in Old Glory, for I know God in heaven flies Old Glory too.

Enjoy the book.

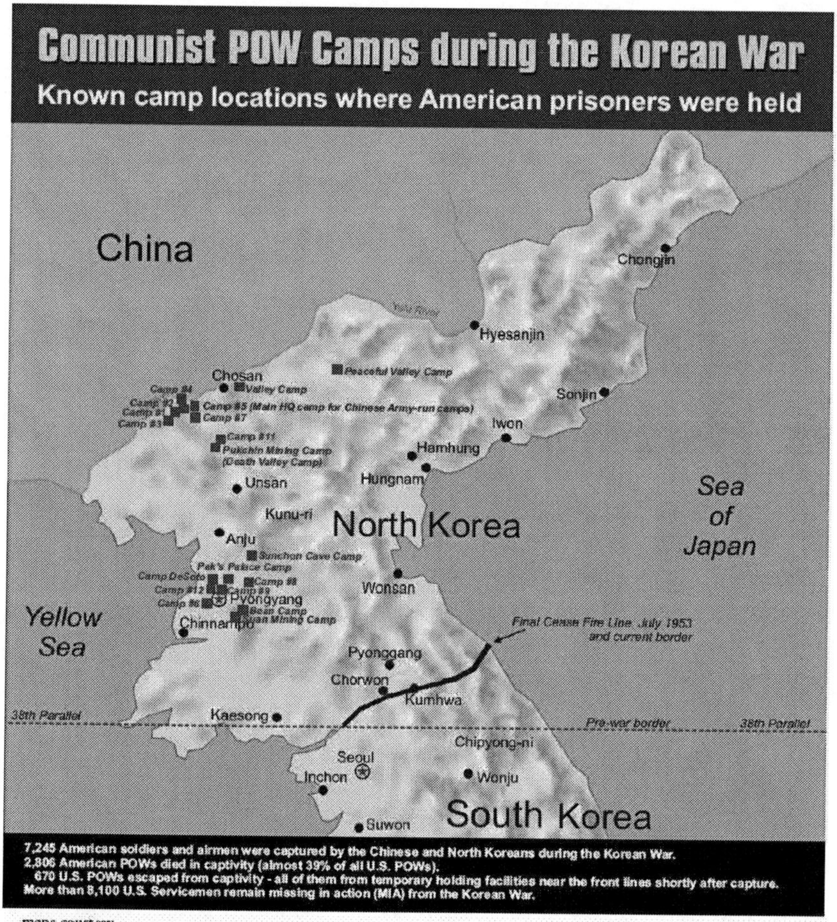

maps courtesy
Geoff Liebrandt
Medical Media Dept
Cincinnati VA Medical Center

The P.O.W. NETWORK
www.pownetwork.org

Acknowledgements

I would like to express my gratitude to all of those who contributed to this book.

First I would like to thank my daughter, Jackie, for the design of our logo.

A special note of appreciation goes to my baby daughter, Marsha, who left home and job and flew a thousand miles to help. Thank you!

I could never have finished the book without the help of my son, Warren, who gave his utmost to me in a very difficult time.

Your old Dad is grateful to each one of you, and loves you so much.

I am indebted to Ralph and Grace Baker, and to my wife, Bernice, for the soundtrack music on the recording of *Old Glory: God Flies Her Too*.

Special thanks to the World Liberty Foundation publishing team and all the others who helped with this project. Their persistence and dedication was a true blessing. I'm also grateful to Ms. Thyme, our media expert. My appreciation for her work is immeasurable.

There were so many good people who provided us with information and stepped up to help whenever and wherever it was needed: My grandson, William; and Dawn and Gary Francois, and their daughter Sharmaine. My dear Sharmaine: what would I have done without you?

There are just so many to thank, but I would be remiss if I did not mention the families of those close, dear friends who gave their last full measure to this great country: the Harper family, the Rybolt family, the Col. McClain family, and the George Pryor family. May they all find peace in God's arms.

Now I come to my dear friend, Col. John Zitzelberger: to express my true appreciation for Col. Zitzelberger would require far more than my ability to convey with pen and paper. Suffice it to say, Colonel, in my view you have always stood head and shoulders above the crowd. Thousands of cold, injured troops should be on their hands and knees today, thanking you for your efforts on their behalf.

As for my final thanks, I must thank my copy editor, Dolores Woodrum. My dear, dear Dolores—her dedication and perseverance to this work is exceeded only by her God-given beauty and talent. If God ever gave one of His children too much of a blessing, it would be Dolores.

I thank all of you from the bottom of my heart. God bless you all!

United States Units in Korea

Eighth Army & Major Units

US Navy

US Coast Guard

Military Advisory Group

A Few Facts About "The Forgotten War"

33,741 US Dead

23,615 Killed In Action

92,134 US Wounded**

4,820 US Missing In Action (Declared Dead)

7,245 Prisoners Of War

2,847 Died in POW Camp

389 POWs known to have been alive after all U.S. POWs supposedly returned

***Lightly wounded who were treated and released back to duty are not included in these figures.*

(04/17/2001 Statistics)

US Air Force

US Marine Corps

CIA

Partisans/ Special Ops.

CIA http://www.korean-war.com/USUnits.html

Prologue

The Korean conflict was a byproduct of the Second World War, which was an amalgamation of two separate regional conflicts. One of them began in Asia in 1937 between China and Japan. Although the two countries had fought intermittently since September of 1931, when the Imperial Japanese Army invaded Manchuria, full-scale war started in earnest in 1937. The other regional conflict began in Europe when Germany invaded Poland under Adolf Hitler in 1939, beginning World War II.

The world's major nations split into two opposing military alliances: the Axis powers, led by Germany, Italy, and the Empire of Japan, considered the aggressors of the conflict. The Allies, led in the European theater by the United Kingdom and France, were joined by the Soviet Union in June 1941 and by the United States in December 1941 after Japan attacked Pearl Harbor. In the Asia-Pacific theater, the Allies were led by the Republic of China.

Within the ranks of the Allied powers, the Union of Soviet Socialist Republics, the United States of

America, and the United Kingdom were known as the "Big Three." U.S. President Franklin D. Roosevelt (FDR) referred to the Big Three and China as the "Four Policemen." In December of 1941, Roosevelt devised the name "United Nations" for the Allies and established the first U.N. charter for world peace. This was based, I believe, on a socialist agenda created by well-known Soviet spies within the Roosevelt administration, including Harry Hopkins and Harry Dexter White. In my opinion, Roosevelt's accommodation of socialism, both domestically and internationally, and his support of Stalin in particular, was the primary catalyst of the Korean war. Had Roosevelt not supported the socialist cause and deliberately made it possible for Stalin to be involved in plans to liberate Korea from the Japanese, the Korean War most likely would never have taken place.

When World War II began, Korea was regarded by the Allies as a victim of, not a party to, Japanese aggression. In November of 1943, three of the Allies met in Cairo to discuss ways to end Japan's occupation of Korea. The Allies who met at the Cairo Conference included British Prime Minister Winston Churchill, Generalissimo Chiang Kai-shek

In June 1945, Harry Hopkins returning from his special mission to the Soviet Union where he met with Joseph Stalin for the last time. Four decades later Oleg Gordievsky, a former KGB agent, stated that Hopkins had been a Soviet agent. Hopkins was the second most powerful man in the United States, FDR's closest aide.

Harry Dexter White, the notorious Soviet spy, helped conceive the Morgenthau plan. White died in 1948, three days after appearing before the House Unamerican Activities Committee.

of the Republic of China, and U.S. President Franklin D. Roosevelt. The first discussions focused on ways to dismantle the Japanese empire. It was at this conference that the participants declared that "...in due course, Korea shall become free and independent." However, privately, Roosevelt could not tolerate having his good and trusted friend, Josef Stalin, being denied access to the Far East-occupied territories and wanted a way to include Stalin in the plans as a major player. He knew that his plan to include Russia would be difficult and would take shrewd diplomacy and maneuvering. A major obstacle in his plan was the non-aggression pact that was in effect between Russia and Japan. During the war Great Britain and the Republic of China came to distrust both Roosevelt and Stalin. They knew Stalin's ruthless history as a butcher of his own people and they believed that Russia should stay in Russia. Great Britain and the Republic of China knew there could be no free and independent Korea with Stalin and FDR involved in the co-occupation.

True to form, at Cairo President Roosevelt introduced his plan of a trusteeship for Korea, whereby the United States, Great Britain, the Republic of China, and the Soviet Union would jointly govern the country of Korea. The plan sounded good, but it had one major flaw—it could not work. With a coalition of four countries that all had equal authority, three of which were free market republics, and the fourth a socialized, controlled society, the plan was doomed to failure from its inception.

The Yalta Conference, held in Crimea in February 1945, was a wartime meeting between Roosevelt, Churchill, and Stalin. By this time the friction and distrust between Churchill and Roosevelt had reached a stalemate. Churchill feared that the close relationship between Roosevelt and Stalin would destroy Britain as an industrial and military power and give Stalin control over all of Europe. It was apparent that Roosevelt wanted Russia to be involved in all post-war decisions involving Korea. Given Roosevelt's statement to William C. Bullitt, the first U.S. ambassador to the Soviet Union, "...I think that if I give him [Stalin] everything that I possibly can and ask nothing from him in return, *noblesse oblige*, he won't try to annex anything and will work for a world of democracy and peace,"[1] Churchill was justified in his concern about the possibility of Stalin taking over western Europe.

At the Yalta Conference, Roosevelt brought up the issue of Korea again, this time proposing a trusteeship involving the United States, the Republic of China, and Russia, for a period of "twenty to thirty years," excluding Great Britain and Churchill. It was at this point in time, in my opinion, that Roosevelt and Stalin came to the agreement that after World War II they would support Communist Mao Tse-Tung over a free and democratic China, thereby putting all of the mainland of Asia under Communist control.

Roosevelt's death in April of 1945 did not allow him to see his plans come to full fruition. The time between his death and the Japanese surrender on August 14, 1945 saw the American–Soviet relationship begin to change. Unlike Roosevelt, who actively supported the Communist movement in America, Truman was passive. The American Congress began to unravel the dictatorial policies of Roosevelt, and the pro-American conservative began to come forward and challenge Truman's decisions.

There was major disagreement between the American Congress and the Truman administration about Soviet occupation policies in Europe. There was also concern about the intent of the Russian entry into the war in Asia. Russia entered the war in the final days of the war just long enough to secure warm-water ports and the productive fishing grounds of Japan's northernmost islands and positioned troops in Manchuria and Korea.

With the Russian entry into Korea at the end of World War II the U.S. State Department, under President Harry S Truman, proposed a plan to the United Nations that would divide the peninsula in half. This plan had the Soviets occupying the industrial, mineral-rich part of Korea north of the 38th parallel, and the United States occupying the agricultural south.

At the Moscow Conference in December 1945, the United States, the Soviet Union and Great Britain again proposed a united trusteeship of Korea and created a Soviet-American Commission with the purpose of establishing a provisional Korean democratic government. No agreement was reached, however, because Russia blocked any proposal that would unite Korea through a free democratic election.

The United States tried to again resolve the issue with the fledgling United Nations in 1947 when the United Nations resolved that Korea would hold elections and elect one national assembly, thus uniting the whole country. The South Koreans participated and elected a National Assembly, with Dr. Syngman Rhee as president of the Republic of Korea. The Russians prohibited an election in the north and denied the U.N. entry into North Korea. Instead, in late 1948, North Korea, with Russian guidance, held unsupervised elections and established the (Communist) Democratic People's Republic of Korea. The North Koreans elected the only candidate running—a hand-picked, Russian-trained officer, Kim Il Sung—as its new president.

In the middle of 1949, with the elections complete in South Korea, the United States wanted to show confidence in the newly elected South Korean government ~~of~~ by withdrawing the American occupation force of fifty-thousand troops to Japan. After the withdrawal, the Americans left approximately five hundred men as a Korean Military Advisory Group (KMAG) to train the Korean security forces.

Roosevelt's successor, Harry S Truman, feared that any escalation of hostilities in Korea would lead to conflict between the United States and Russia. Truman's plan involved using the KMAG to train the South Korean army, not as a military force, but as a police force equipped with light artillery, light tanks and small arms. That hardly equipped Syngman Rhee to defend against a North Korean attack or launch an attack on the North Koreans. Attempting to reunite the peninsula by South Korean military action was, therefore, out of the question. The North Koreans, being influenced by Russia, had no preconceived idea of a unified, freely-elected and independent Korea. Kim Il-Sung, however, had convinced Stalin and Mao Tse-tung that he could unite the Korean peninsula in one day. The Soviet Union supplied the 150,000-man North Korean army with heavy tanks and long-range artillery that was far superior to the poorly supplied rag-tag police force of South Korea, leaving South Korea vulnerable to attack from the North.

The year 1949 brought about the desertion of the Republic of China by its allies which determined the fate of China to be controlled by a Communist dictatorship. The withdrawal of new equipment and logistical support from the Republic of China by the U.S. State Department assured a Communist victory over a free democratic China. The Chinese Communists under Mao Tse-tung won the civil war and forced the military troops of the Republic of China to flee to the island of Formosa, also known as Taiwan.

This success by the Chinese Communists, combined with the U.S. position toward the Far East, allowed Kim Il Sung in 1950 to gain support for the uniting of Korea by force with the blessing of Mao Tse-tung and Stalin. The U.S. position was to contain Communism on mainland China and not allow the Communists to displace Chiang Kai-Shek from Formosa. At this point in the war, however, the U.S. issued a statement that its defense of Asia did not include Korea.

On June 25, 1950, the North Korean forces attacked South Korea across the 38th parallel in an attempt to unite the peninsula under Communist rule. During the five years between World War II and the Korean War, Truman had gutted the standing U.S. Army and Marines to the point that they were totally unprepared to enter Korea, or any other conflict. Nonetheless, Truman committed the

U.S. forces that were in Japan to defend South Korea and repel the North Korean forces.

The North Korean act of aggression south of the 38th parallel caused the United Nations to authorize for the first time a multinational force under the U.N. banner to "repel the aggression." The U.N. resolution called for the immediate cessation of hostilities and the immediate withdrawal of all North Korean Forces back to the northern side of the 38th parallel. The resolution passed for two reasons: Russia, as a Security Council member, would not vote to fight against itself. Also, the Russians were boycotting the U.N. because of its recognition and seating of the Republic of China (now in Formosa).

The U.N. Security Council passed a second resolution asking its members to furnish "assistance to repel the armed attack and to restore international peace and security to the area." Under another resolution the United States was asked to lead the police action, and President Truman reluctantly accepted the request and appointed General Douglas MacArthur as commanding General. Twenty other countries contributed to the effort, but the majority of the conflict fell onto the American solider.

The initial North Korean offensive overran the poorly trained and poorly equipped ROK forces. The ROK army, after being reinforced by American troops, finally stopped the advance of the North Korean army and was able to establish a fifty-mile-wide defensive position on the southern tip of the Korean peninsula, called the Pusan Perimeter. After heavy losses the conflict reached a stalemate.

General MacArthur devised a strategic yet risky plan to cut the over-stretched supply lines of the North Korean army and, in one bold stroke with the amphibious landing at Inchon, retake Seoul, the capital of South Korea. MacArthur's troops quickly surrounded the North Korean army and destroyed its fighting ability. Eighty-five thousand North Koreans surrendered, and the remaining stragglers withdrew to north of the $38^{th.}$ Parallel.

Seeing the opportunity to unite Korea and repel Communism, General MacArthur proposed to advance the U.S. Army toward the North Korean/China border at the Yalu River. Truman, fearing escalation of the war by the entry of the Chinese Communists, disagreed with MacArthur's strategy. Nevertheless, MacArthur persuaded

Truman to allow him to push north to the Yalu River, although with severely limited resources.

As the American forces under MacArthur approached the Yalu River and the Chinese border, Truman, still fearing escalation by the entry of the Chinese Communists, caused indecisiveness in the American forces. MacArthur was told at one point to bomb only half a bridge and not follow the retreating North Korean forces across the border, which left a sanctuary and an assembly area for the Chinese troops to attack on the other side of the Yalu River.

MacArthur was defeated by the Chinese Communists at the Chosin Reservoir and was forced to retreat back south of the 38th parallel.

The long-running frictions between Truman and MacArthur resonated throughout the entire Far East Command. These frictions led the Communists to maximize and exploit the situation through their propaganda machine in the U.S. and the worldwide Communist movement. This friction, to the delight of Communist China and North Korea, led to the replacement of a commanding general in the field while fully engaged in combat. Truman rewarded MacArthur's defeat at the Chosin by replacing him with General Matthew Ridgeway, who was in complete lockstep with Truman's containment strategy. The departure of MacArthur and the placement of General Ridgeway as commander added more confusion and indecisiveness to the situation, but General Ridgeway was a soldier's soldier and a stabilizing factor in the situation.

Mao Tse-tung, the little-known leader of now Communist China, encouraged by Russia, with the desire to be recognized as a new, powerful world leader and someone to be reckoned with—a status he could not accomplish as a simple guerrilla fighter—needed an action that would propel him to the forefront of the world's stage. The opportunity he needed was presented to him in Korea.

In November of 1950, the massive Chinese forces and elements of the defeated North Korean army now assembled at the Chinese border and launched a massive attack at the Chosin Reservoir, entering China into the war and giving Mao Tse-tung the arena he needed for the promotion and recognition of his version of the Communist movement.

The Chinese forces overwhelmed the U.N. forces by superior manpower. The vastly outnumbered and ill-equipped U.S. Army retreated from fighting the Chinese. They were not prepared for the bitter cold that had encompassed the Korean peninsula. Both sides sustained severe losses, and the U.N. forces withdrew to a line just north of Fusan, but by the spring of 1951 they had fought their way back across South Korea to the Han River and the 38th parallel.

The Chinese attempted several offensives with huge losses, but by June of 1951 the battle lines were set along the 38th parallel. Unable to accomplish their objectives, the Russians submitted a cease-fire to the U.N. and requested negotiations.

For two years the peace talks continued while U.N. forces kept fighting at tremendous loss of life. In 1953 the United States inaugurated a new president—a soldier named Dwight David Eisenhower—and by July of 1953 an armistice agreement was signed by all but South Korea, which did not want to accept a divided Korea.

To date there has been no political settlement and no official end to the Korean War. Only a stalemate exists between Communist forces and the United Nations.

Could the Korean War have been prevented? The answer is yes. However, the only man who could have prevented that war was FDR. He was the only one of the Allies at the Cairo Conference in 1943 that had been willing to overlook Stalin's butchery when he was killing thousands of his own people. Churchill and the China were not.

Roosevelt's friendship with Stalin blinded him to the realities of the times. Stalin was determined to see a free China turn Communist and the whole of Europe under Russian control, leaving England without an industrial complex or an army, reducing it to no more than an agricultural economy. By 1949 China had fallen Communist. Mao Tse-tung had taken over the mainland of China, forcing the leaders of the Republic of China to flee to the island of Formosa, causing President Truman's Secretary of State, Dean Atchison, to remark "We have sold out the Chinese people."

Harry Truman was the last president to be in a position to prevent the war, but he failed to do so. He let it be known that Korea was of little interest to the U.S., convincing Kim Il-Sung (the Russian puppet), Mao Tse-tung, and Stalin that America would not defend Korea. The world watched as Truman ripped the U.S. Army apart and tore our ability to defend our special interests apart at the seams.

Russia had exploited Truman's weakness for appeasement rather than victory over Communism and had blocked off most of Eastern Europe with the Berlin Wall. Truman's reply to the Russians taking of Europe was an airlift of food to the starving people in Eastern Europe. Russia continued to build the Wall and reinforce its position and the puppet government in North Korea. The North Korean army had been built to a 135,000-man fighting force that would attack south and take South Korea, thus giving Russia a foothold in Korea, extending its land grab that had begun at the Cairo Conference in 1943.

The fighting ended when the American people elected General Dwight Eisenhower as President. This let the Communists know that the American people now were ready to take the war to the next level rather than allow the slaughter to continue.

In later years, Truman's frustration with trying to fulfill the wishes of Roosevelt concerning Communism and Stalin, and his obligations to the American people, were written in his diary, but it was never conveyed to the American people nor to Stalin and the Russian people. As to the Korean

War, he recorded these questions to Stalin in his diary, "Do you want an end to hostilities in Korea, or do you want China and Siberia destroyed? You may have one or the other. Whichever you want. You either accept our fair and just peace offer or you will be completely destroyed." This was typical of Truman never to publicly convey his true feelings toward the Russians.

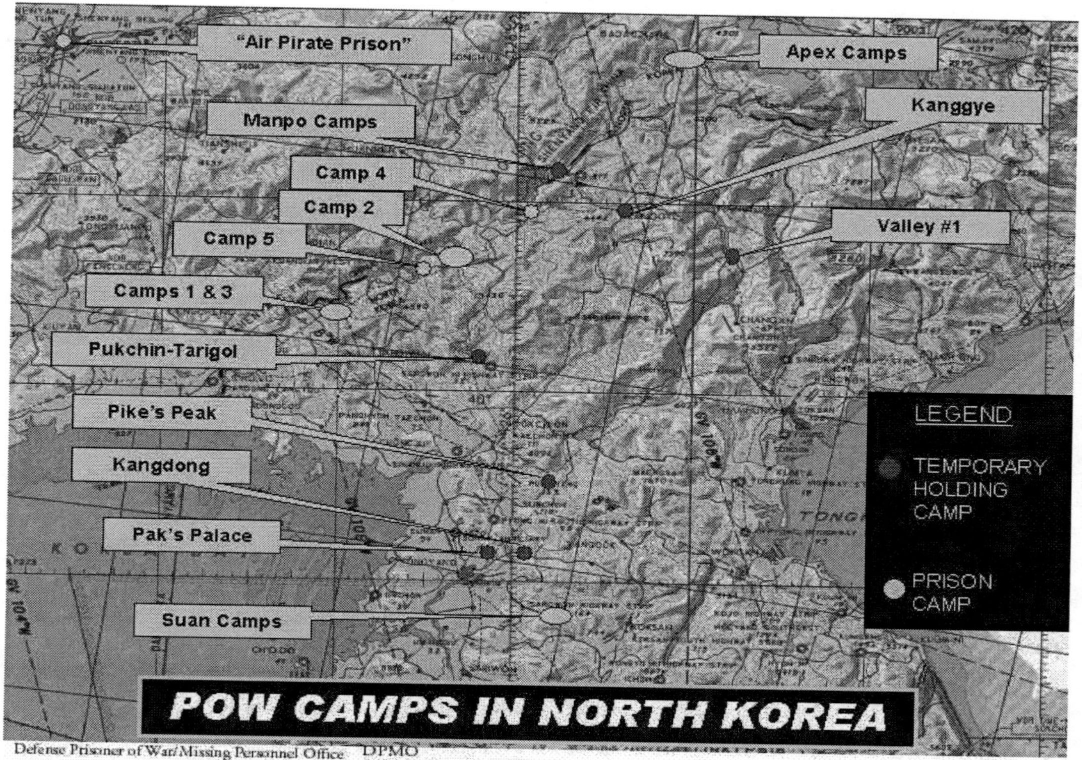

Defense Prisoner of War/Missing Personnel Office DPMO
2400 Defense Pentagon,
Washington, DC 20301-2400

Chapter 1
Korean Occupation, 1948

It was an early January morning in 1948; the smell of "honey buckets"—buckets carried on ox carts that picked up raw sewage from the night before—filled the air. The whole city of Inchon smelled like one large cesspool. I was eager to board the train heading north. My orders were for an Army camp near the 38th parallel.

We boarded the train around noon for the sixty-mile ride to an old Japanese military camp. The camp was made up of huts heated with diesel fuel; large fifty-gallon barrels lined the exterior of each hut and the barracks. I soon learned that when the temperature went below zero the barrels froze and we were left without heat.

I was assigned to the communications platoon. The man in the bunk next to mine was a switchboard operator. He was leaving in two weeks—his time was up and he was going home. I went to work with him the following morning.

The switchboard was located in the building used as battalion headquarters. It was clean and warm. “Man,” I thought. “This is where I want to be!”

I arranged a meeting with my new boss, a very young-looking second lieutenant whom I soon discovered was an outstanding communications officer.

Lieutenant Barnes was a small man who worked hard, right along with the other men. He was always ready to share information, no matter the insignificance of the question, and he always made you feel good about asking him anything. He gave me two weeks to learn to operate the switchboard. I never looked back.

Lieutenant Barnes knew I meant no disrespect toward him so I was never reluctant to ask him to give us a hand. Our relationship grew into one of the most rewarding experiences of my Army career.

Two months after I arrived, he walked into the communications hut, handed me buck sergeant stripes, and told me I was his communications chief.

“Lieutenant,” I said, “I’ll give it my best shot; but I don’t know a thing in the world about being a communications chief.”

“Yes you do, Abbott. I’ve watched you with your nose in the training manuals. You can handle it, all right. At any rate, my communications chief is going home and you're all I've got."

"You have my sympathy, Lieutenant." We laughed.

Lieutenant Barnes called a meeting of the communications platoon that morning and introduced me as the new communications chief. He brought in another man to run the switchboard to free me up to check the lines for breaks and disruptions between battalion headquarters and the outposts we were responsible for. He asked me to drive him to one of the outposts stationed along the 38th parallel.

For the next two weeks we followed our telephone lines to the next outpost. Lieutenant Barnes explained to me how crucial it was that our lines remain in good condition.

Our machine guns were pointed north, and the Russians' machine guns were pointed south. At that point in time no North Koreans were stationed along the border. Lieutenant Barnes explained the political situation, but I found it difficult to understand how the Russians and the Americans could have been allies during World War II and now be pointing machine guns at each other. I knew nothing about

Communism. I was soon to learn Russian Communists were real.

Soon things began to happen. Our lines were tapped. Korean conversations were heard on our lines. I found myself traveling in a Jeep through the mountains, often by myself, following our lines. Strange faces began to show up in the villages. The village where I had always stopped to have tea no longer made me feel welcome. Kim, my interpreter, no longer felt safe in my Jeep. Previously she would ride with me all day. She worked for the South Korean Syngman Rhee government.

Kim was well educated and spoke perfect English. She was North Korean, and the most beautiful woman I had ever seen. Her family had been forced to give up their home in Pyongyang and move south past the 38th parallel. She was part of the exodus of two million people forced to leave North Korea and settle in the south. She would not discuss her father's occupation. He appeared to do nothing but live well. I could always count on him for a bowl of rice, a cup of tea, and an occasional cup of hot sake. Conversation with his daughter was always enlightening.

Local Koreans with whom I got along well provided the unskilled labor in and around the camp. Our doors were always open, which was an open invitation to would-be thieves, and they began to steal everything that was not nailed down.

Once our battalion went on field maneuvers for two weeks. When we returned we found that every footlocker in the communications platoon had been ransacked and was empty—except mine. It had not been touched. Our platoon sergeant and sergeant major reported the loss to the company commander.

"The only clothing the platoon had left was on their backs, except for Abbott. His footlocker was untouched," they said.

"Now, Abbott," the first sergeant said, "we have a serious problem."

"Oh? Sergeant, what might that be?" I asked.

"Did you know while we were in the field the barracks were ransacked? The contents of all of the footlockers in the communications platoon were stolen except for those in your footlocker, Abbott." His voice got a little louder. "Would you like to tell me about that, Abbott?"

"Tell you about what, Sergeant?"

"Don't you think it is a little strange that every locker was ransacked except yours? How do you account for that, Abbott?"

"I don't know, Sergeant. Count my blessings, I guess."

The first sergeant, Sergeant Swanson, was a crusty old devil and a no-nonsense soldier. He had seen me in downtown Seoul during monsoon season. I was getting tired of being wet, so I bought a large umbrella. I was walking down the street, happy and dry, when I heard footsteps coming behind me, then his voice:

"Abbott, is that you?"

"Yes, Sergeant."

"What the hell are you doing with an umbrella? Don't you know that soldiers don't carry umbrellas?"

"Sergeant, just because a man is in uniform surely does not mean he's lost his mind, does it?" Then I said, "Seriously, Sergeant, I was not aware it was against regulations. I apologize for that."

"Get rid of the damned umbrella."

"Okay, Sergeant." I folded up the umbrella and handed it to a Korean woman who was passing by.

"Good," he said.

The old sergeant was in a Class A uniform and looked like a drowned rat. I tried to contain my emotions, but could not. I started laughing uncontrollably.

"What the hell's so funny, Abbott?" The brim of his hat was filled with water and was overflowing onto his shoulders and face.

"Forgive me, Sergeant, but you look like a man in full dress uniform under water. Are you sure you don't want me to catch up with that lady and get my umbrella back?"

He shook his head, looked me square in the eye, and said, "Soldier, you're about to learn that breaking military rules is no laughing matter. Be in my office first thing in the morning."

"Okay, Sergeant," I said.

I reported to the sergeant the following morning. He gave me five days CQ (in charge of quarters). I apologized for poking fun at him.

I knew there was no way I was going to pull CQ. The rain was destroying our telephone lines to the outpost. I related the story to the battalion commander. He called the company commander and the first sergeant.

The first sergeant said, “Forget it; he’s learned his lesson.” Then he said, “Abbott, do you still own an umbrella?”

I said, “No, Sergeant,” and he smiled

The first sergeant was from the old school, all business. Anything taking place in the company was his business. He looked after the men like a hen with little chicks. This was his company, all others be damned.

We never found out who stole the clothing from the communications platoon.

One day the first sergeant came by the communications shack looking for me. He wanted to talk. He was going back to the States soon. He and I talked about the increase in gang activities. The sergeant was showing his soft underbelly. He was concerned about my safety while running lines by myself through the mountains. He encouraged me to take someone along, so from then on I did whenever I could find someone who was available to go.

Lieutenant Barnes was being transferred to the Philippines. He received permission to take me with him. The battalion commander objected to my being transferred, stating I was needed here.

I was supposed to be promoted; however, President Truman was decreasing the size of the Army and had frozen the ranks. I would not get another stripe but would remain communications chief.

The open lines built by the Japanese became unreliable because they were easily tapped. A crew of linemen from the 7th Signal Corps arrived to help us run new closed lines to the outpost; they were all good men and within three weeks we had new lines and a reliable backup.

Late summer approached and we received communications replacements. They were good men, well schooled in communications, but they were not equipped to handle the problems we were facing—seeking and finding wire taps in the mountains. The situation was serious enough for the battalion commander to furnish me a Jeep on a twenty-four-hour basis.

The new motor pool officer was hell-bent on deadlining vehicles on schedule. He deadlined my Jeep and refused to give me another one, stating that all available vehicles were past due for maintenance and none were available. That lasted three hours. The battalion commander needed to contact the officer at Outpost 22 and was unable to get through on the phone line. When I informed him my Jeep had been deadlined, he had the motor pool officer bring me his personal Jeep at two a.m. so I could go clear the lines.

Ten miles from camp the line left the valley and went out over the mountain. Some of the poles we had set on top of the hill were dangerous to climb. We set them in solid rock but we did not have the right equipment to set them deep enough. I was sure no one would even attempt to climb the poles and tap the wire. I was wrong.

I stopped by to see Kim's father in the village. I brought candy bars and cigarettes for him. Kim was there and motioned for me to follow her into the courtyard. She told me to be careful; four men were climbing the mountains to get to the phone lines. One man had a roll of wire and the other had a telephone like the one I was carrying. I asked where these men were. She said she did not know, but thought they had gone back down to the road.

I thanked her and ran back to the Jeep, drove quickly down into the valley, and checked the line to the switchboard. It was clear. The line to the outpost rang, but there was no answer. There was no sound to indicate a tap but my gut feeling told me that there was a tap, and it was close. I drove around the mountain about two miles and checked the line on that side of the mountain. It was clear to the outpost.

There was no time to go back to camp or go back down off the mountain and call for support. It would soon be dark. I had to get out of the mountains. I found an overhanging rock, pulled off the road, removed the coil wires from the Jeep and put them in my pocket so no one could steal the Jeep.

I headed up the hill and got lucky. The tap was in the first pole, about fifty feet from the road. I climbed the pole and cleared the tap. From the top of the pole, I could see the road and valley below. There was no one in sight. I cut the tap.

As I looked down, I could see where they had dug a trench to bury the wire. A poncho covered what appeared to be a foxhole. I pulled the wire, and there was a telephone connected to it. I got down and picked up the phone. Then I climbed six feet back up the pole where I could see the road below. In the distance I saw four Koreans dressed in white. They were about three hundred yards from me. I climbed down the pole and ran with the telephone back to the Jeep, but it would not start. Then I remembered pulling the coil wire off. I took it out of my pocket and put it back on the Jeep, started it up, and headed back to camp.

The following day I went back to check the line and noticed a phone line hanging from the pole. When I stopped, three Korean soldiers came from out of nowhere. Evidently they had been hiding and waiting for me. One of the Koreans tried to grab my weapon. He had both hands on my carbine but was unable to pull it out of my hand since it was wedged in the steering wheel. I hit him in the mouth with my right hand and he fell like a rock. I drove off at high speed. I had trouble stopping the bleeding that was coming from my hand.

In the rearview mirror I could see the Koreans running full speed toward me, waving their hands for me to come back. This would be the last time I would try to run the lines without someone riding shotgun for me.

When I stopped the Jeep, I could see the impact of the soldier's teeth had separated my pinkie finger from my right hand. It took thirty days in the hospital for my hand to heal.

During my stay in the hospital I shared nightly chats with the head nurse. Upon discharge she asked if I wanted to return to duty.

"I have no choice," I said.

"I can tell you how to fool these doctors, Abbott."

"Oh Captain, a man cannot cheat his way through life. One must live with himself and do what he can to carry his load, regardless of the consequences."

I had spoken with the communications officer in charge of camp utility; he would have no part of a transfer. "What would you do with an easy life, Abbott? You belong in the thick of things, out on the 38th parallel."

"Have you no compassion, Lieutenant? If not for me, then what about my poor mother?"

"What's your mother got to do with this, Abbott?"

"Well, Sir, I know if I told her that her favorite son was out of harm's way, doing a great humanitarian service, helping the sick in the hospital, she would sleep like a baby."

He laughed and said, "Get the hell out of here, Abbott, and go back to where you're needed. I've spoken to your commanding officer." As I walked away he said, "Abbott, we need someone with a sense of humor around here. If I have an opening, I might try to get you transferred."

"Thank you, Lieutenant," I said, and that was that.

I was sitting on the side of my bed, thinking about what I would do when I got back to duty, when Sergeant Jackson, one of the nurses' assistants walked in.

"I hear you're going back to duty on Monday, Abbott," he said.

"Yes," I said.

He asked, "How would you like to go on a picnic at the river?"

"That sounds great, Sergeant, but I have no clothing. It's locked up."

"I'll get clothing for you."

"Okay; I'll see you tomorrow morning around seven-thirty."

"Great," he said.

The following morning I was up early, shaved, showered, and ready to go. Sergeant Jackson brought my clothes at six-thirty and said he would be back in two hours. I got dressed and walked across the compound to get a cup of coffee from the canteen. The first person I saw, as I walked through the door, was the nursing supervisor. She was sitting alone at the table next to the door. She invited me to sit down. She did not ask how I had gotten my uniform. She inquired about my hand, and I said it was healed.

"You're going back to duty Monday morning."

"Yes, Captain, I am. I'm going to miss you, Captain, and our nightly chats. You have given me an education. I would like to stay here a little longer, but I know I'm needed back at the company."

I turned to look out the window and saw the truck pulling up. Then I turned toward the captain, and said "Captain, would it be proper for me to visit you when I get a weekend off?"

She laughed. "Yes Abbott, I'd like that. Thank you."

When I got on the truck, I could see the captain sitting at the table, and I thought, "Abbott, the road to hell is paved with good intentions. You have seen that lady for the last time. And that's probably how it should be. Enlisted men have no damned business getting a crush on a commissioned officer. Nothing good can come of that." I never saw her again.

We stopped in front of a house in Seoul. Sergeant Jackson got out, and within five minutes he returned, accompanied by four Korean girls. We loaded up and headed to the river.

It was a warm summer day. The drive down the dirt road to the river was relaxing. The land surrounding the river was still except for an occasional farmer working in the field. The road ended and turned into an ox trail. The driver put the truck into four-wheel drive and we drove another mile up the river where we stopped for lunch.

The beach was more rock than sand but the water was clean and clear, and sufficient for swimming. The water temperature was surprisingly warm. Jackson suggested we fan out along the bank and gather some firewood while he prepared lunch. We gathered wood and built a fire. Jackson had thought of everything. He brought Spam, pork and beans, Blue Ribbon beer, and fresh baked bread. Jackson missed his calling; he should have been a chef. At that time beans and fresh baked bread was a feast fit for a king.

After chow, we washed our gear and stowed it back on the truck. Sergeant Jackson and the two orderlies who had come with us jumped into the deep water with their girlfriends, leaving Hy and me standing there.

Up to that point I was not sure how this was going to work. Now I knew. Hy suggested we walk up the river away from the others. That suited me just fine; I did not question her motives. I was soon to learn she did not have a bathing suit, and she did not want to swim in front of the other folks in her birthday suit. As for me, I had no intentions of wearing anything.

Hy was seventeen years old, in full bloom, and was an experienced, gracious Geisha. I thought I had died and gone to Heaven that afternoon.

The sun was going down when we got dressed and started back to join the others. Jackson hollered for us to get on the truck. Everyone was on the truck except Hy and me.

We were about a hundred feet from the truck when out of nowhere six young North Koreans, gang boys, started throwing rocks at the truck. Hy and I began running as fast as we could over the rocky river bed.

The Koreans were attempting to cut us off from reaching the truck. I told Hy to keep going as I picked up two baseball-sized rocks. An older Korean boy said, “I don’t want to hurt you, Shorty, I just want the girl.” I ran past him. Hy was safely on the truck and the truck was moving. I looked over my shoulder just as the Korean threw a rock. I jumped to one side and the rock hit the tailgate. I turned around and saw the little bastard reaching down to pick up another rock. I threw my rock with everything I had; my pitching arm was in good shape that day. I hit him in the ribs and he fell down and rolled over, screaming in pain. I jumped on the truck.

A rock hit the driver, causing him to lose control of the truck temporarily, and the truck hit a large rock in the road. Hy was thrown across the back of the truck and hit her mouth against a bench, which knocked out four of her teeth. I gave her my t-shirt to put over her mouth.

After we had gotten back on the main road we raised the side curtains to let in some fresh air. Hy pointed toward the front of the truck. One of the North Koreans had managed to get into the truck and hide. He may have gotten in while we were still in the river. I had no way of knowing.

Jackson called back, “Is everyone okay back there?”

"Yes, Sergeant, but what the hell are we going to do with the North Korean who's sitting in the front of the truck, showing his teeth like a fat cat?"

Sergeant Jackson stopped the truck, changed drivers, and got in the back with us. We headed down the one-lane road, going twenty-five to thirty miles per hour.

Jackson grabbed the North Korean by the collar and literally threw him into the back of the truck. The little Korean said nothing. He was just standing there, looking out the back of the truck, when suddenly Hy walked to the back of the truck and pushed him out.

"He would not jump," she said.

No one spoke a word during the remainder of the trip back to Seoul.

We stopped and dropped the girls off at the house where we had picked them up. Hy did not say a word to Jackson. He packed her mouth, gave her a shot for the pain, and told her he would return the following day to dress her mouth. They dropped me off at the rear entrance to the hospital, and I went straight to bed.

The next morning I went down to the recreation room and played cards until late that evening.

Monday morning I was up at five a.m. and waited impatiently until eight a.m. for the doctor to release me, which left me with one hour to catch the train north to the 38th parallel and Camp Page in Chuncheon, South Korea.

I arrived at camp and went straight to the communications shack. Sergeant Major saw me and motioned for me to come to his office. He said, "The colonel wants to see you."

I reported to the colonel and he said, "Abbott, I need you to find out what's happening to our phone lines. Get to the bottom of the trouble and fix it."

"I'll get right on it, Colonel," I said.

As I drove back to the communications shack to find someone to ride shotgun with me, I saw one of the Koreans working outside the barracks. The thought occurred to me, "Hell, I need to know what these people are saying on the telephone," so I asked him if he wanted to ride with me. I explained to him that it might be dangerous and he might get shot.

"No, you will not let me get shot," he laughed.

Just on the outskirts of town I climbed a pole, tapped the line, and handed the phone to my Korean helper. He laughed and said, "Laundry...laundry!"

"What laundry?"

"You can hear," he said, and handed me the phone. I'd never heard such a racket in all my life.

He knew where every laundry was in town. We followed the line back and sure enough, at our second stop, we found the problem.

The . Korean who owned the laundry did not know how the phone got there. I asked if he talked on the phone, and he said "No." He did not know who the person was who was using the phone. He said he had recently fired a man because he would not work. I asked him if he was familiar with the Syngman Rhee police. He said he was.

I asked him, "Do you want them to burn your laundry down, beat the hell out of you and put you and your family in jail for a long time?"

The old man began to cry and begged me not to call the police. "I am not a spy," he said.

I took the phone; that cleared the line and everyone was happy. Whoever was on the other end of that line with another tap disconnected his phone. Finding him would have to be for another day.

We went back to camp. I gave my Korean helper a carton of Lucky Strike cigarettes and dropped him off, and then I went and notified the switchboard operator to let the colonel know the lines were clear.

The company clerk came out of the order room and told me that I had patrol duty that night.

"Damn, Corporal, have I been missed that much? What did y'all do when I was in the hospital?"

"First Sergeant said you are going to do your part to make up for the time you lost from the company while you were in the hospital."

"Corporal, tell the old shit I love him too."

Lieutenant Rounds was the duty officer. I learned a great deal from Lieutenant Rounds; he taught me to control riots. It didn't matter to him if a mob gathered around his Jeep, he would stand up in the Jeep with a stick in his hand and tell the driver to drive forward at full speed. Then he would bust heads. It was not pretty, but it was effective.

When it came to riot control, Lieutenant Rounds would say, "Abbott, take the first one, the most vocal…he's the one who is trying to get things started and will stand back to let others do the damage. Let him know that you know who he is and what he's up to; make sure he understands that he will pay the price. The blood and shit splattered on the walls will be his."

Once on a trip to Seoul by train we had a long tunnel to go through. There was no light in the tunnel. There was a young lady who was with the USO sitting beside Lieutenant Rounds. She remarked that she was afraid of the tunnel and of the dark. Rounds put his arm around her and reassured her that she was going to be all right. About halfway through the tunnel, she yelled, "Rounds, get your damned hand off my leg!" When we came out of the tunnel, Lieutenant Rounds was sitting there with his hands folded and looking straight ahead, like the cat that ate the canary. He looked at me and winked, and I whispered, "Some folks are just damned unappreciative, Lieutenant." He didn't say a word, and we never spoke of the incident again.

Guerilla activities increased. Once when Lieutenant Rounds was the officer on duty we went uptown on a routine security check and found ourselves surrounded by a mob. One of the guards said, "Abbott, this Korean civilian tried to take my weapon away from me."

He pointed to a well-dressed young man. All eyes were on him. It was a tense moment. I told my men to load their weapons and stand up in the Jeep.

"If I give the order to shoot, don't take aim, just shoot and kill as many people as you can."

I was watching the well-dressed Korean to see what his reaction was when I told the men to shoot indiscriminately. I could see he had started to retreat. I told the man he tried to take the carbine from to push his carbine straight out, where the young Korean could reach it.

"When he touches it," I said, "I'll kill him." I said it loud enough so all could hear me. Most Koreans could understand and speak some English.

The guard pushed the carbine into the man's chest. I had my .45 pointed between his eyes at a distance of three feet. He took one look at me as I pulled the hammer back on that .45 and then put his hands over his head and walked away. That broke up the mob.

Syngman Rhee's army was supposed to be sufficient to repel a North Korean attack. In truth, the South Korean army was nothing more than an undisciplined police force.

Rhee was constantly on the radio bragging about the strength of his police force. His police had their own cruel way of obtaining information from individuals. One morning I walked by a barbershop. The sign on the door said it was closed but I could see movement within the shop. There were four of Rhee's finest men standing around a man who was sitting in the barber's chair. They were spinning him around, each one slapping him as he passed. It was a bloody mess.

If you're smart, you don't see things like that. I did not mention what I had seen to anyone. From that point on, when I saw one of Rhee's finest, I stayed clear of him.

One of the Koreans who worked in the camp told me that Syngman Rhee's police were bad people. They were gang members, he said. We had gang members among our gate guards.

Within a few months the whole attitude among the South Korean population seemed to change due to the massive influx of North Koreans being sent south under the guise of overpopulation.. Fraternization among the U.S. armed forces and

Korean women was prohibited. A Korean woman caught with an American by these gangs was dragged through the streets so everyone in the village could see. I once saw two of our gate guards, along with two other people, dragging a young girl through the mud with her hands and feet tied. Her crime was talking to an American soldier.

The Syngman Rhee government asserted that the Korean people were tired of occupation, and that was creating unrest. This was untrue. The Korean people knew nothing but occupation. They were always under somebody's control, including Japan's. The Japanese occupation, they felt, was good for them. You would be hard-pressed to find an older Korean who would criticize the Japanese occupation.

Older Koreans were reluctant to discuss the Japanese occupation, but when pressed by Americans would criticize the Japanese because they thought that was what the Americans wanted to hear. In reality they knew the Japanese had given them roads, railroads, electricity, prosperity, and good schools. As one Korean told me, the Korean population was well taught by the Japanese. "The Japanese taught us to read and write in both Korean and Japanese," he said—and with a great deal of pride, not resentment.

The American troops were leaving South Korea, and the South Koreans could not understand why we were leaving. I asked one of the Koreans whom I had come to know quite well, "What will you do when we leave?"

He said, "I will go Pusan. My mother and father have laundry business in Pusan and I can do well there. Kim Il-Sung," and he pointed north, "no go Pusan. You leave, he come here."

"Well, don't you think Syngman Rhee will protect you?"

"I not think so."

We closed our business day and radio down at 22:00 hours (eight p.m.). I would go to the radio shack and listen to Radio Moscow which had continuous music, interrupted only by propaganda messages by North Korean President Kim Il-Sung. Kim Il-Sung would come on the air with the same message, "We will liberate our South Korean brothers and sisters and crush the imperialist puppet, Syngman Rhee."

Syngman Rhee, on the other hand, was on the local radio stations assuring the population that his police force could repel and crush any attack from the north. Both North and South Korea broadcast in both Korean and English.

Why were both the north and south wanting to make sure the Americans got the message? Syngman Rhee was no longer worried about his defense and was telling the Americans they no longer needed to defend his country and his homeland—Korea could defend itself.

We left the 38th parallel and, after a brief stop in Seoul, we moved on to the Inchon Harbor and went aboard ship to go to Japan to join the occupation forces defending the northern islands of Japan.

While on the fantail of the ship in the harbor, looking back at shore, a second lieutenant and I were standing, shoulder to shoulder, and talking.

"How secure, do you feel, is South Korea?" I asked the lieutenant.

"The south can hold its own," he said, "against any attack from the north."

He was shocked when I said I felt preparations were being made for an invasion. Two million refugees arrived in the south for economic reasons. But my experience in the villages along the 38th parallel led me to believe that part of the exodus of two million people to south Korea was intended to act as an advance invasion force, perhaps to set up a supply line to protect the invasion when it came, and to provide a safe haven for guerilla thugs to disrupt the Rhee government.

He said his sense was that Rhee was pretty secure. I pointed to the beautiful police academy building sitting up on the hill overlooking the harbor. "When we come back, Lieutenant, that building and all these other landmarks will be gone. For us, Lieutenant, it's either San Francisco in '48 or a salt mine in '49."

"I hope not," he said.

Bill Abbott, 1950

Chapter 2
To Japan, Camp Crawford

After we landed in Japan we traveled across the country by train through the cities where the atomic bomb had been dropped. There was total destruction.

Two inches of snow covered the ground. Little children ran alongside the train, wearing only diapers, and would scream and beg for food. Whole families were living in cardboard boxes or whatever shelter they could find to stay out of the cold.

I noticed a complete section of rail sitting on top of a blown-out building; it was twisted like a pretzel. What an awesome force of destruction the bomb was!

Our destination was Camp Crawford in northern Japan, on the island of Hokkaido. The barracks were clean, with hot and cold running water, and heated with steam. The PX was loaded with anything you had the money to buy.

I walked into the dial exchange and saw my first cousin, Sergeant Billy King. I was to be his replacement when the 11th Airborne returned to the United States. "What a small world," I thought.

When the 11th Airborne returned to the United States it had discarded many of its undisciplined troops, and the 7th Division became a dumping ground for deadbeats. Some of these men formed gangs and were nothing but troublemakers. They had no communications skills and did not appear to want to learn any.

Lieutenant Miley took over as company commander. He was an outstanding West Point officer and wanted me to replace First Sergeant Newman, an old World War II veteran who drank too much.

This was Christmas, 1949.

Several of my men rented a cab and were going to a party. Some Japanese girls had invited Private First Class Bresondean to a party about fifteen miles north of Sapporo. I was unaware that the whole city was off limits to GIs.

The cab driver called ahead and asked if I was there, and if the soldiers had permission to go to town. I assured him the CID (Civilian Intelligence Department) had given us permission to go to town.

The club was first class and had a beautiful dance floor. The drinks were cheap, and lots of free beer and fresh oysters were set on each table as a

Christmas gift to us. We were the owner's guests of honor. The Japanese band was outstanding. We sang Christmas carols the whole evening, until closing time.

I was thanking Bresondean for inviting me when suddenly the place was full of Japanese police. They placed everyone under arrest. One of our men had hit a Japanese man and insulted his wife. We were taken to the Japanese prison.

I convinced the police captain my men would cause no more trouble, so he gave me permission to call a cab and return to camp. Suddenly the same man who had caused all the trouble in the first place started running through the prison, pushing guards aside, grabbing keys and unlocking doors, screaming like a wild man. More guards showed up in riot gear and moved in. I asked the captain if I could try to stop the man before the guards moved in and somebody really got hurt. He told the guards to let me try.

The man came running down the stairs. He had gotten a nightstick from somewhere and was hitting the stair rails.

"The riot police are here," I said "and someone is going to get killed. Why don't you give up your stick so we can go on back to camp?"

"Somebody *is* going to get killed, and I'm going to do the killing."

"You're not going to kill anyone."

"And you're going to stop me, Abbott?"

"If I have to. You're my responsibility."

I picked up a wooden chair and broke it on the stairs, leaving only a leg of the chair intact. He lunged toward me and I hit him with the chair leg. He used his left arm to block the blow. I had swung hard enough to break his arm but hit him on the head, and he fell over the stair rail. That ended the prison riot and we went back to camp.

Due to the prison riot my men were all put under house arrest. The company commander, Lieutenant Miley, ordered me to take my men on a five-mile forced march as punishment. I was happy to do that and hoped that would be the end of it.

Those men were ungrateful bastards. They wanted to cheat on the march. One mile out, one fellow said, "Let's just sit here a while and then go back."

"You damned people will never learn," I replied. "You're facing stockade time and you want to play games! We are going to complete this five-mile march. I don't want to hear any more about it." I didn't know at the time that Lieutenant Miley had sent a man to follow us.

A special and general court-martial was in the works. These men had bad records. They were sentenced together and given dishonorable discharges. Private First Class Thompson and I were tried separately. We were found guilty and sentenced to six months' confinement and forfeiture of all pay and allowances, and a bad-conduct discharge. The court asked if I had anything to say.

"Gentlemen," I said, "I'll do whatever you deem appropriate and forfeit my pay, but I cannot live with the disgrace to my family, country, and myself of being separated from the Army under less than honorable conditions. Let me make it clear that I am not speaking on behalf of Private First Class Thompson. What I am about to suggest has not been discussed with Private First Class Thompson. In this situation, he must speak for himself. I would like this honorable court of distinguished gentlemen to consider the following: let the disgrace that I have brought to my country end today. Let my life end in recognition of payment for the perpetuation of that disgrace. It is my request to this honorable court that these charges be rescinded and I be tried by general court-martial and sentenced to death by firing squad."

The court fell silent. No one spoke a word. They just glanced at each other and lowered their heads. After what seemed to be about five minutes, Private First Class Thompson and I were marched under guard to the stockade.

The stockade at Crawford was a simple fenced-in compound. It was composed of one large unheated tent. Each prisoner was given a fold-up cot, a sleeping bag, a mess kit and one change of clothing. The bathhouse was an unheated wooden shed. To reach the bathhouse/latrine you had to walk a hundred feet through the snow.

Looking back, the stockade at Crawford was primitive but it served the purpose of punishment. It was a reminder of how good it was to be back in the company. I believe it should have been the model for the entire Army. A soldier should know the price there is to pay for his unlawful actions.

Thompson said, "You crazy SOB, I'm sure glad you told them you weren't speaking for me. I ain't giving my life for this honor bullshit. I'll handle the bad conduct discharge; it won't bother me one bit." Thompson told the other men in the stockade what had happened in court.

"Thompson," I said, "you keep your damned mouth shut."

Thompson did not say another word. The other prisoners just looked at me, shook their heads, and walked away in disbelief.

The court ordered a review of the case, rescinded the discharge, and the regimental commander sent word to me through the non-commissioned officer in charge of the stockade that he wanted me to be transferred to regimental headquarters. I sent word to the colonel that I was indebted to the officers and men in the 3rd Battalion Headquarters Company. I would return there and try to make amends.

General Barr, the 7th Division commander, closed the stockade, releasing all of the prisoners. I went back to the company, which proved to be a mistake because the men I was going back to help were no longer there.

Upon arrival at the company, Lieutenant Rybolt, the communications officer, invited me to have coffee with him. He said, "I'll get your stripes back."

I said, "I appreciate that, Lieutenant, but with a black mark on my record it's too soon to get a promotion. I'll be going back to the States very shortly. Find a good man and I'll help you train him. My best bet is to play it safe, stay clear of trouble until it's time for my discharge, and then go back to the farm where I belong. In the meantime, I'll do everything in my power to help you."

I had no idea the Korean War was going to break out. All communications-trained men would be shipped to Korea, leaving the company with nothing but new faces and untrained men. The communications school was closed just when it was desperately needed.

The regimental communications school was opened in 1949 with two instructors, a staff sergeant and myself, when we arrived from Korea. The staff sergeant transferred, leaving me as the only instructor. I held classes in the field and in the classroom.

I was very proud of that platoon. I had a strong, deep-rooted belief that the most important part of an infantry unit is communications. There is no substitute for the ability to communicate.

Communications Platoon
Camp Crawford 1949
Photo courtesy of the family of Jerry Francois

Chapter 3
To the Inchon Invasion

We received our orders to move out. Our platoon was loading equipment on the train for the trip south to the Inchon landing. I drove a three-quarter-ton radio truck to the railhead and was told it would be six hours before we could load the truck onto the train. I went back to camp, loaded my personal belongings, and headed to town to say goodbye to my good Japanese friends.

I had a large turntable and an amplifier that I wanted to give to my Japanese friend, Joy, and her five-year-old daughter. Her husband had been killed in World War II. I became close friends with Joy and her father. Her father was a postmaster who loved horses. We spent many wonderful hours at the racetrack together. He knew horses and would win considerable amounts of money—as I did when he told me what to bet on.

As we said our goodbyes, Joy's daughter just stood there with her lips poked out, tears flowing down her face. I took her in my arms and told her mother to take her daughter to her grandfather, away from the GIs, and give her a chance for a good life.

As I hugged and kissed Joy goodbye, the radio operator called my name. It was my turn to load the truck.

I headed for the railhead.

"Don't cry," I said to Lee. "I'll be back."

"No," her mother said, "You will not be back. Good soldiers don't come back. They die in battle. And Abbott, you're a good soldier."

As I boarded the train, Joy gave me a bottle of sake as a going-away present. That gift would cost me dearly—it would cost me my master sergeant's stripes.

Our troops were riding in regular passenger cars, which required sleeping sitting up. I went back to the radio truck, placed my sleeping bag on the floor, got in it and went to sleep.

First Sergeant reported that I was not on the train. Lieutenant Rybolt assured him that both Greer and I had loaded the trucks, and sooner or later I would show up.

Later I walked back through the passenger cars, saw nothing but long tired faces, used the bathroom, and headed back to the radio truck.

The next day was a beautiful day. The sun was shining and the air smelled fresh and clean. I found a Jeep with the top down. I got out my bottle of sake, sat down, and was enjoying a sip of my going-away present when Captain Snodgrass came along. He wanted to know why I was not with the other men. I invited the old captain to sit down and enjoy the ride with me.

"Captain," I said, "Let's enjoy this beautiful Japanese countryside together."

"Why aren't you riding with the other men, Abbott?"

"I can't do it, Captain Snodgrass. I look at their faces and it bothers me. These men have no training. They're like sheep going to the slaughter. Captain, are you familiar with the Korean way of life?"

"No."

"They have a total disregard for human life and a deep-rooted hatred for Americans. I'm going to enjoy this beautiful scenery for perhaps the last time, Captain Snodgrass."

The old captain was amiable, until he looked down between the seats and saw the bottle of sake.

"What's this bottle, Abbott?"

I said, "This is a going-away present, Captain, from a dear friend. I'm enjoying a little nip; it's good for the soul. Would you like to share a drink?"

The old captain became enraged. He called me a drunk, saying I was unfit for duty. The truth is, Captain Snodgrass had never met me before that day.

"You are a professional drunk, Abbott," he said.

"Wait just a damned minute, Captain! You don't even know me. I assure you, I'll be ready for duty in the morning. It's you, Sir, I'm concerned about. Captain Snodgrass, I can blame my stupidity on this bottle of sake. But you, Sir, what, pray tell, is your excuse?"

The captain got out of the Jeep and walked away. I had wanted to talk to Captain Snodgrass about Korea and what we could expect from the Korean people, but when he saw the sake bottle, the opportunity for conversation was gone.

My mind wandered to the reality I did not want to face. For the last few hours my thoughts had been on the good times…spending time at the racetrack with Joy and her father…weekends at the beach…the good times at Camp Crawford, and the pleasure I received from teaching at the communications school. I remembered the graduation party we had at the school. That was the first time I had tasted champagne. I drank it like it was water and got sick as a dog.

Those were the days when I believed the officers and men were an extended family. That was now gone. Captain Snodgrass' actions ended my belief in our extended family.

There was no point in getting frustrated when someone like Captain Snodgrass did not want to hear what I had to say. It only made matters worse for me. I should have humored him and thrown away the bottle of sake.

We were going to war. My job was to stay alive and do my job as a lineman and assume no further responsibility. I needed no rank—my discharge date was less than a month away. I would be careful not to be classified as essential, which was the case in 1948 when I was denied a transfer to the Philippines. The ideal situation for me would be to have Lieutenant Rybolt assign me several helpers; I would handle installing and maintaining telephone lines and just stay out of the way of other people until the war ended. Captain Snodgrass believed I was a drunk. If that was the prevailing attitude among the officers, I would not be able to make much of a contribution to the war effort.

The battalion arrived and began setting up bivouac at the base of Mt. Fuji to prepare for the Inchon landing. Training began to take place, but it was too little and too late.

I walked over to the training area after checking telephones and installing new batteries. The men were bogged down. It seemed no one knew how to connect the switchboard. They called Sergeant Presky, the wire chief. He saw me coming to the area near the switchboard and told Lieutenant Rybolt, "Abbott can handle that. Let him do it." I showed the men how to connect the switchboard.

Someone called me from company headquarters to inform me our South Korean troops had just arrived from Seoul. Our recruits were straight off the streets of the city, I was told. Someone had been trying to teach these Koreans how to shoot their weapons for two days without success. Captain Snodgrass said, "Who do we have that can teach the Korean troops to shoot their weapons?"

"Give them to Abbott," said Lieutenant Rybolt. "Abbott, what about it?" Lieutenant Rybolt was determined to get me my master sergeant's stripes, but I didn't know it yet.

I said, "OK, Lieutenant, I'll take the troops."

Lieutenant Rybolt said, "The company commander wants to see you."

Captain Snodgrass was still upset from the confrontation we'd had on the train coming south.

"Let's start over, Abbott."

"Yes, Sir. The sake was just a gift, Captain. My Korean friends meant no harm, Sir, and I do not apologize for their actions."

"That is all, Abbott."

I saw him scratch through my name, and that was the last time I spoke with Captain Snodgrass.

Harry Cutting was the acting platoon sergeant and Captain Snodgrass gave him the stripes that were meant for me.

When Lieutenant Rybolt heard what happened, he was mad as hell at me for being such a fool, and was mad with Captain Snodgrass for being so bull-headed. I hated myself for it, but the damage was done. Lieutenant Rybolt let me know he expected me to be responsible for training the communications platoon, and that I would still be responsible for the South Korean soldiers.

"OK, Lieutenant," I said.

"Abbott, you are your own worst enemy."

"Yes, Sir," I replied.

I had no problem teaching the Koreans how to shoot, and I made good wiremen out of them too. Most of them stayed with me until I was captured by the Chinese at the Chosin Reservoir.

We received orders to board ship for the landing at Inchon. After boarding the ship we were waiting for orders to sail.

Some of us were standing on deck, looking out over the dock. I turned to a merchant marine, a member of the ship's crew, and asked him, "How long will it be before we shove off?"

"Oh, five or six hours, I was told."

Private First Class Williams said to me, "A cold beer would sure taste good."

"It sure would," I said.

The sailor said, "You have plenty of time to get one." He pointed to a place about half a mile away.

I had fifty dollars in my pocket and no place to spend it. "Williams," I said, "are you game?"

"Let's go get that beer," he said.

We headed out, found the restaurant, and ordered fried oysters and a large pitcher of beer. We had just ordered the second pitcher of beer when two MPs walked into the restaurant.

"Oh, shit! Look who's coming, Abbott! We don't have a pass."

"Not much we can do about it now, Williams. They would just take us back to the ship," I said. "Enjoy your beer."

The MPs were all business. "What do you think you're doing here?"

"Well, Sergeant, we're killing a little bit of time. We're chowing down on some of the finest oysters in Japan, and we're drinking some cold beer."

"Do you soldiers have a pass?"

"We are not on leave. We're just enjoying some chow and a cold beer before going back to the ship. It sails in about four hours."

"In other words, you're AWOL?" the sergeant asked.

"I suppose, technically, we are AWOL, Sergeant." I paid the bill and started to leave when that ape of an MP shoved his nightstick into Williams' chest and pushed him back into the booth.

"Where do you think you're going?"

"We need to get back to the ship, Sergeant. We're on our way to Korea," I said.

"Look, Soldier, you are AWOL in an off-limits area. We've been picking up deserters like you for the past week. I'm going to take your ass to jail, and until someone from your company comes to get you, you'll stay locked up."

"But you don't understand, Sergeant. We'll be in serious trouble if we miss that ship. The ship is docked down the street. Why don't you take us down there? Someone will verify who we are and we can go on aboard. Everybody will be happy. There's no point in taking us to jail. Let's be realistic; we are members of the invasion force which is on that ship. No, we don't have a pass, but we aren't deserters either. That invasion force is more important than some damned person's ego."

"No, we're going to take you down to the provost marshal and notify your company that you're here."

We were put in the Jeep and taken about five miles from the dock to a holding pen for processing. There must have been fifty or more men waiting to be processed. By the time we were processed, six hours had gone by. I was pleading with anyone who would listen. Our ship was going to leave us. We always got the same answer: “You should’ve stayed aboard.”

We were processed in time to notify the ship we were there. Lieutenant Rybolt and another officer came to pick us up, but before we could get back to the ship it had sailed, leaving us standing on the dock. Just then a full colonel pulled up in a Jeep. When told of our circumstances, he said, “I’ll get you aboard a ship and you can transfer to your unit at sea.”

We boarded the flagship General MacArthur and set sail. Within a short time the other ship and the convoys were in sight. There was no conversation between the four of us. We all knew we were in deep trouble, but we would be in deeper trouble if we could not join our unit for the invasion.

Lieutenant Rybolt and the other lieutenant stood on deck with us and said nothing. I knew in my heart that they would be court-martialed, and it was my fault.

Lieutenant Rybolt and I moved to one side and started discussing the coming events. This would be the last extensive conversation we would have until Pusan.

As we moved off the ship and climbed over the side to join our unit for the invasion, Lieutenant Rybolt said, “Abbott, some of us will not be coming back.”

“You’ll be all right, Lieutenant...I’m deeply indebted to you, Sir. I’ll not soon forget that.”

We arrived at Inchon harbor and were waiting for high tide. The invasion forces were ready to set in. Our Navy was shelling the city of Seoul, thirty miles away, with big guns that sounded like trains as they passed overhead.

Suddenly a North Korean plane flew over our deck. Anti-aircraft guns opened up. All our ships were trying to hit this one little airplane without success. The plane circled back and headed toward our hospital ship.

The hospital ship was as big as a football field and was lit up like a Christmas tree. An Australian ship was anchored near the hospital ship. After all the attempts by our Navy to hit the little Russian-manufactured airplane, it only took one shot from the Australian ship to down that Korean plane.

Someone said, “I’m glad those Aussies are on our side!”

“You got that right,” I said.

We went in about the fourth wave of the invasion. We landed without serious resistance. There was some small-arms fire on our left flank, but it didn't affect us.

As we landed, I heard someone calling my name. It was First Lieutenant Miley, my old company commander.

Chapter 4
Inchon to Pusan

That evening I met with Lieutenant Miley for first time since the war had broken out. Lieutenant Miley was the company commander of the 187th Infantry Airborne Regimental Combat Team. He needed a platoon sergeant. Tech sergeant stripes were being held on his desk for me. He had something else on his desk: a fifth of twelve-year-old Old Charter whiskey.

"Let's have a drink, Abbott. We're going to jump north of Seoul at 0500 hours this morning and cut the enemy escape route back to the north."

Lieutenant Miley pointed out the area in which the jump was to take place. It was an area I knew well. I had spent a month in the hospital two hundred yards from their destination. I had worked on the telephone lines in that area and picked apples in the orchard a hundred yards south of their drop zone.

"Lieutenant," I said, "I tried to find you to transfer to your unit. You are the one officer in the

Army I would like to fight this war with. You are a credit to the officer's corps, Sir."

"Abbott, I'll put you on the jump roster and we'll jump together in the morning."

"I would like nothing better, Lieutenant Miley, but you know the old adage: I'm caught between the devil and the deep blue sea. I want to serve you, but at the same time, I'm indebted to Lieutenant Rybolt and he needs me desperately. He's in over his head and without help. I must give my loyalty to him. But I do have a few suggestions for you, Sir, on your jump in the morning."

I pointed out that the rice paddies north of the tracks were wet.

"Our intelligence tells us they're dry, Abbott."

"What they mean, Sir, is they've done an aerial reconnaissance and there is no standing water in the paddies, which is true. They have been drained, but the paddies are nothing but one big mud hole. If your equipment is dropped there, you will lose it. My suggestion is to move south of the railroad tracks and drop into the apple orchard. Then attack north along the tracks. North of the tracks it gets hilly, and it's rough. It would be difficult to assemble your men there, even without enemy resistance.

Enemy resistance on the hillside would be very difficult." Then I said, "Good luck, Lieutenant."

I headed back to our bivouac area. It was late but I couldn't sleep. I kept thinking about Lieutenant Miley and his men. "If the Koreans have the high ground and are in fact dug in, there will be a bloodbath of vast proportions. He might get lucky, however, and land his troops southwest of the rice paddies. If he does that, he'll be okay," I thought.

Lieutenant Miley was an outstanding West Point officer, the kind of man you can count on. I asked God to walk with him as he and his men jumped that morning.

I hoped I had made the right decision. One thing I knew for sure—Lieutenant Rybolt was in over his head and he desperately needed help. Sergeant Kresky was the only knowledgeable communications NCO he had, and he had been transferred to a line company, leaving only me. Now I was all he had. I would see to it that his switchboard and phone lines were installed and maintained. That would make him look good.

Sergeant Maynard, the radio chief, would follow orders and keep his radio section operational.

Sergeant Cutting, the new communications chief and platoon sergeant, knew little or nothing about the responsibilities of a communications platoon. On top of that, he was on the lazy side.

Then there was Private First Class Greer, who was hard-working and loyal. I could count on him, and loyalty is very, very important. I knew he would take care of things for me. I would make him wire chief. He could obtain the supplies we needed. Greer was also a good rifle shot, and that was extremely important. Sooner or later we would be in a serious firefight and I would need him desperately.

There were a few other good men in the company I could count on when needed. Greer and the five ROK (Republic of Korea) soldiers assigned to me in Japan were my complete support. The soldiers had to be trained. I started training them immediately. One of them was a professional boxer who spoke English well. He would be in charge of the other Koreans. Once everyone understood what was expected, everything fell into place.

I called a private meeting with my Korean soldiers. I told them, "This is your leader," pointing to the professional boxer. "His name is 'Syngman Rhee,' after the president of your country. If you have a problem, you tell him and he will tell me and I'll fix it. I will teach you how to be a lineman. I expect you to work hard every day. We start work at daybreak and we stop when the job is finished. You will not work for anyone unless I give you permission to do so. If anyone asks you to work for them, you tell them to talk to me. I'll get you clothing and shoes that fit. I'll feed you, and you will be treated like the American soldiers. When we encounter enemy forces, you will respond with deadly force. If anyone feels he cannot kill his northern brother, I want you to step out now. I'll release you. But if you stay and we come in contact with the enemy, you must fight as a soldier. If you do not fight like a soldier, I'll shoot you myself. Every man must be confident we can depend on each other. I will not ask you to do any job I will not do myself. The last thing I have to say is the most important: what we do, how we do it, and when we do it is not to be discussed with anyone. There will be no discussion of these matters outside this group. Any leak of information could get you killed. We work as a unit, eat as a unit, and fight as a unit. Sergeant Syng will move you to the mess hall and we will have chow. In one hour, we start to work."

Enemy resistance was nonexistent except for a few pockets of North Korean troops trying to escape to the north. The Inchon invasion had been a success beyond anyone's wildest dreams. Seoul fell within days, though there was serious fighting in the areas around it. It was clear to me, as I listened to the radio reports and the information I got from tapping into the telephone lines, that the North Korean army was in full retreat. There was no longer an organized fighting force in South Korea.

The 3rd Battalion set up in a small village. The communications shack was in a schoolhouse. The perimeter was spread out. Some companies were two, three, or five miles apart, causing a nightmare for our line crew. But we were up to the task.

It was my policy to stay as far away from the communications platoon as possible, going there only to receive orders from Lieutenant Rybolt.

These were difficult times. I pushed my men hard from dawn to dusk. We were unable to eat in the mess hall while the other men were enjoying hot meals, so my men were eating cold C-rations.

Greer would work hard, and when we stopped to eat he would bury his nose in a Zane Grey western book. He reminded me of a little boy with a comic book. Every time we stopped and I wanted to discuss something that needed to be done, I had to tell him, "Put the book down, Greer."

One day we worked from five in the morning until nine that night, stopping only long enough to eat a can of beans. The house had electricity—one sixty-watt bulb in the middle of the building. Greer was lying directly under the light, reading one of his westerns.

"Turn off the light, Greer. We have a big day tomorrow." I dozed off, and when I turned over, the light hit me and woke me up. Forgetting where I was, I saw Greer sitting there, still reading. "Greer, turn off that light. What in Heaven's name is wrong with you, man? These men have to get their sleep. Turn off the damned light and go to sleep!"

After about five minutes, I said, "Greer, I'm not going to tell you again to turn out that light. If you don't turn it out, I'm going to shoot it out. So keep reading if you want to."

Greer continued to read the book as though nothing had been said to him.

A few minutes later I said, "Gentlemen, cover your eyes." I picked up a .45 and shot out the light.

"Greer," I said, "I'll deal with you in the morning. I'm going to make a switchboard operator out of you yet. Get your shit together or I'll have Jerry François transferred here to take your place tomorrow."

From that day forward Greer came to me every evening, asking what we needed to do the following day to get started.

Several days later we had a full crew of men. We pulled into a village comprised of a group of about five houses, and saw a Russian one-and-a-half-ton truck. It looked like it was brand new. The occupants were not there. Their weapons were left in the back of the truck. There were three WWI '03 rifles.

We searched the village for North Korean soldiers and found none, but I was sure the locals were hiding them someplace.

We looked the truck over. I decided to take it and use it to haul material. I asked Jerry François to drive the truck. François was a talented fellow who could be counted on when you needed him.

We'd gone about two miles with the enemy truck when suddenly, about one hundred yards ahead, I could see artillery being fired at us, blowing big holes in the road. The forward observer had spotted an enemy truck coming down the road.

We stopped the truck. I climbed the nearest pole, tapped the line, and got straight through to battalion headquarters. "I've captured an enemy truck," I said. "Would you please hold your fire?"

They quickly stopped firing.

That Russian truck rode like a dream. It was head and shoulders above the American Army's rough-riding trucks. It had an air compressor to inflate tires, two twenty-five-gallon gas tanks, two spare tires, a wench, and three metal racks. That truck was ideal for what we needed.

I went to the company commander to get rations. He refused, saying the line companies were short on rations and we could not have any. "The line companies come first, Abbott." I thanked him and went on my way. I was on my best behavior. I was not going to make him mad, as I had done with Captain Snodgrass, which had cost me my master sergeant stripes. It was difficult not to tell him he wasn't talking to an idiot. I knew that every line company in the battalion had their kitchens in full operation and every man had access to hot food.

I had brought my men in from the field the night before, and the only thing we had to eat was apple pie and canned milk. We made a meal out of the pies. I thanked the mess sergeant for them and told him I would not bother him again. From here on in, my men would eat C-rations as long as the war lasted.

The following morning, just before dawn, I cranked up the Russian truck and headed for the dock at Inchon. American soldiers were unloading rations and placing them on the dock. I knew the soldier assigned to guard duty; I knew he was the one to account for the rations that were placed on the dock. I would not ask him for rations. The rations being unloaded had been designated to be shipped to the 187th Airborne. I would have to get my rations directly from the ship.

"Since I have to hang around until they start unloading rations for the 7th Division, our division, I'll relieve you for a few minutes, Corporal, so you can go over there and get a cup of coffee and a free doughnut, compliments of the Red Cross." He handed me his clipboard and headed out.

When he got out of sight, I backed up the truck and yelled to the crane operator to give me twenty-four cases. He dropped them in the back of the truck, and I pulled the truck back from the dock.

I never counted the cases, but there were more than twenty-four. Whatever the number, they were sufficient to carry us all the way to the Chosin Reservoir.

The guard came back with his doughnuts and cup of coffee and thanked me for relieving him.

We drove off, stopped for a few doughnuts and coffee ourselves, and headed back to work. We loaded the truck with wire, telephone and climbing equipment, put a tarp over it, assigned a place for rifles and ammunition, and fixed a place for the men to ride.

"Gentlemen," I said, "this is our home for the duration of the war. Get your gear, including your sleeping bags, put it on the truck, and fill your canteens with water. We have cigarettes and C-rations. We will not depend upon the mess hall for food. If we are near a mess hall at chow time, we'll take advantage of a hot meal. Sergeant Syng, make sure we have a hibachi and wood to fire it up."

I was always in contact with Lieutenant Rybolt. He gave me orders every night as to what should be done that day. I loved Rybolt like a brother. I teased him, but there wasn't anything I wouldn't do for him.

As we worked our way south, there was little or no enemy resistance. Thousands of refugees were moving out of Seoul, heading south. One morning we were trying to run a line along the road and the string of refugees walking along the road was so thick we were unable to get to the poles to hang wire. I noticed an unusually well-dressed Korean woman, perhaps in her late twenties. (Korean women seemed to look young for their age.) I told Syng I wanted to talk to her.

She said she was coming from Seoul, her family was in Inchon, and she was going to stay there until the fighting in Seoul was over. She was college-educated and had worked at the officers' club. She had not eaten in two days. The North Koreans, moving north, had taken all of her family's food.

I told Syng, "Give her three cans of rations; one for her, and one each for her mother and father." She didn't want to take the rations because she said she could not hide them. Other refugees would take them away from her and her parents. I told Syng, "Walk with her and her parents up the hill and have them stay in the house where we stayed last night, and tonight we'll feed them."

We fed the woman and her parents. They spent the night and left the following morning. I told Syng to tell them they could stay in the house. Our soldiers had vacated the house and would not be back. The woman said her mother needed to see a doctor and they must travel to Inchon.

That was the way it was: thousands and thousands of South Koreans were displaced from their homes and looking for safety, carrying everything they owned on their person. Many were like this young lady, trying to get a sick loved one to a doctor. Many of the Korean doctors were helping the American medical staff.

At Inchon, in an effort to get communications established as soon as possible, I had utilized existing open lines. The Japanese had installed them prior to the war and they were still standing. One morning I climbed a pole to tap a line and used it to connect two outposts, when suddenly the glass insulator shattered, and a large piece of glass became embedded in my head. At the aid station the Korean doctor put six stitches in my head.

We talked about the devastation of his country, and the old man cried. The Korean people knew what it was like to be occupied by foreign troops but had never witnessed such total devastation of their country and national spirit.

The Koreans are a proud, family-oriented people. They saw their families torn apart by their neighbors from North Korea across the 38th parallel, and now by the Americans. The man on the street, like the doctor, wondered if it was all worth it.

The 7th Division moved south toward Pusan, a seaport city on the southern tip of the peninsula. Pusan had been the site of furious fighting with great losses for both the Americans and Koreans. It was the last stronghold of the American troops, with their backs to the sea. The successful invasion at Inchon, which cut the enemy forces in half, had freed Pusan from the stranglehold of the North Korean forces.

In war, one of the most beautiful sights is the surrender of enemy troops. We were witnessing the surrender of thousands of North Korean Communist troops. The surrenders seemed to indicate more and more that the war was over. It was now a formality to let the politicians announce the end of the war.

There were, of course, pockets of resistance, but nothing serious. We encountered a group of die-hard North Korean forces by accident. They were members of a guerilla force which was using hit-and-run tactics, hiding in the hills, shooting troops that were riding on the outside of tanks as they went through deep passes in the mountains. Narrow roads with high banks provided maximum cover for the attacking forces firing down on the tanks below.

The troops riding in open trucks and on tanks were like sitting ducks. Tanks are useless under these conditions and circumstances. A tank's inability to fire overhead makes it vulnerable to attack from above.

We stopped in a little village that appeared undisturbed. We had chow, posted a guard, and went to sleep. Just before dawn, the mayor of this little town walked up and approached the guard. The guard awakened Syng, who woke me up and told me the mayor wanted to talk to me.

The mayor said there was a rifle platoon of North Koreans about three hundred yards east of our position. The North Korean troops were cleaning their weapons and getting ready to relieve the other half of the platoon, which was guarding the pass.

They were killing those riding on tanks coming through the narrow pass. He said they had raided the village of food supplies and his people were facing starvation.

I had four GIs and five South Koreans. The mayor said there were about thirty Communist soldiers, maybe less; he did not know how many there were, and he did not know how many were at the pass waiting for the tank column to come through.

I said to Syng, "Tell him I'll send a patrol to look at the situation."

Then I said to my troops, "Gentlemen, I leave it to you. I would do this diplomatically, but since I'm short on diplomacy today, get your shit together and let's go! This could be a trap. We all could be wiped out, or we might catch them by surprise. Regardless of the situation, there will be a firefight and you will earn your Truman beans this day. Now, we do have another option—we could just drive on and let the tanks fare for themselves and let the town handle the Communists the best way they know how."

"Syng," I said, "what is your feeling? Is the mayor telling the truth?"

"I believe him, Abbott."

"Well, Syng, have one of your men stay here with our equipment, and keep the mayor with him. If he's not telling the truth, I'll shoot him when I return. Get an extra clip of ammunition and at least two grenades apiece."

The mayor of the town was right. Just as we reached the top of the hill overlooking the small village, we saw that there were four houses: three overlooking the valley on the south side and one house on the north side, just as he had said. An old Korean woman was sweeping the yard of one of the houses on the south side. The mayor had described it down to the last detail.

The sun was coming up over the hill, reflecting off something shiny that seemed to be penetrating the top of the house, near the east corner. Upon close observation I recognized it was a machine gun. I called Syng over and he pointed to the west corner. It also had a machine gun, as did the other two corners. □□

We worked our way off the hill, getting out of range of the machine guns. We gathered at the porch of the house. I couldn't believe the North Koreans had not posted a guard. I told the men to crawl around the house and position themselves in front of each door. □□

"When I give the signal, throw all the grenades you have through the doors. And throw them hard enough to make sure they break through the door. We don't want any bouncing back in our laps." □□

Upon my signal, it went off like clockwork. All you could hear was grenades exploding and men screaming with pain. I called for surrender and walked over to the haystack that was in the yard, when suddenly a North Korean soldier hit the ground in front of me.

Syng said, "Here's your guard," and handed me a rifle. I instructed Syng to tell him the war was over for him and that he would be well taken care of. If those in the house did not surrender immediately, we would set fire to the house and there would be no surrender, all would die.

The first man coming through the door, with his hands over his head, ran full speed at Jerry François.

François shot him dead. I lost it for a moment.

I told François, "If you shoot another prisoner, I'll blow your head off!"

I pulled him aside and told him, "If these people knew they are surrounded by only ten people they'll fight to the end." (Actually there were not even ten people because three Koreans refused to get into the fight.) "If they choose to do that, we stand a damned good chance of not leaving here alive."

I told Syng to apologize for the soldier being killed.

"Let's start putting hay on the porch. If they don't come out by the time we finish, we'll burn them alive."

One of the ladies in the house on the hill came running down the hill, screaming at Syng and pleading with the men inside the house to come out. She didn't want her house burned down. I told her she couldn't come any closer.

I told Syng, "You tell her I don't want to kill her, but if she comes any closer, I'll shoot."

Suddenly a voice inside the house said six of his men were wounded badly and wanted to surrender, but would only surrender to the commander.

I said, "Syng, tell him I'm the commander and I'll accept his surrender, and assure him that all his men will be treated fairly, and will receive medical treatment and good food."

The captain surrendered. Twelve more men walked out. Two others were dead, including the one François killed.

When the captain surrendered, he had his hands above his head, but he was holding a pistol. Syng walked over and knocked him down and took the pistol out of his hand. Syng then reminded me of an instance when a North Korean captain had removed the clip from his pistol but left one round in the chamber and then raised the weapon over his head to surrender, to show me how simple it would have been for the captain to simply tilt his pistol downwards and shoot me in the head. I thanked him.

"It's an old Japanese army trick," he said. "Abbott, you get killed easy."

We marched the prisoners back to the truck.

The mayor was overjoyed.

Within minutes a column of tanks came into our field of vision. They were approaching around the bend. The tank commander, a first lieutenant, came to me and inquired about my prisoners. He asked if they were part of the unit that was firing on his tank column. I told him they were the relief unit and we had intercepted them before they could relieve the other unit.

"Perhaps the delay in the relief unit caused the other unit to withdraw early. You can never be sure, Lieutenant, what goes through people's minds. Lieutenant, do you have transportation for these prisoners?"

"Yes. We have an empty truck at the rear of the column. We can accommodate them."

"Sergeant, can we take these prisoners off of his hands?"

"Yes."

"Corporal, take charge of these prisoners. "

"Gladly, Sergeant."

The master sergeant pulled the corporal over to one side and said softly, "Now when these gooks get away from town they'll try to escape and our lead tank will open fire on them."

Unaware that I had overheard the conversation, the sergeant apprised the lieutenant of his intentions to do away with the North Korean troops. The first lieutenant walked past me without saying a word.

I called Syng and Greer over to the side and said, "Get ready to take these prisoners back. These bastards plan to kill them, and I won't tolerate that."

I walked over to the lieutenant and said, "Lieutenant, I need to talk with you." We walked about twelve feet away.

"Lieutenant," I said, "let's understand one thing. These are my prisoners, and I'll kill the first man who harms one of them, whether it's your first sergeant, or corporal, or anyone else, yourself included. These men have laid down their arms and surrendered to me. They are my responsibility, and I'm taking them back. I know your men have seen hard times and have lost several men, but killing these unarmed men will not bring one of your men back. There will be no ego trips at the expense of my prisoners, Lieutenant. I'll have Greer and Syng take the prisoners off your hands."

"Greer, line up the prisoners and guard them."

One of the prisoners broke off running across the rice paddy. At least fifteen men from the tank company opened fire on the little Korean without ever hitting him.

"Hold your fire!" I said.

The Korean soldier took cover in the corner of the rice paddy. I walked over to him and told him it would be okay, that I would see to it.

I knew he spoke English because he was the only one near enough to overhear my conversation with the lieutenant. He thought he was going to be shot anyway. □□

"Get up and come with me. I'll not let anyone harm you, okay?"

"Okay," he said.

"Now, Son, if these men were better shots you would already be dead, and that would have served their purpose and it would have been a waste. If you try to run again, I'll be the one doing the shooting and you will not be able to hide in the rice paddy. Now get up and let's go join the others. Your comrades are over there waiting for you."

He put his hands over his head and jumped up, I thought to run again. I shot him with the commander's new eight-millimeter, Russian-made pistol, hitting him in the left temple.

He was dead. I rolled him over. He had a rock in his hand.

I told his commander, "You should decorate that man, for he was a good soldier."

"Greer, lead these prisoners on over to the truck and let's get ready to move out."

I was glad to get away from the truck and tank commander. They had seen serious fighting before the Inchon landing broke the enemy's back and it was beginning to show. It's strange how combat affects a man. As he kills the enemy, as he sees the enemy kill his comrades around him, it has a devastating effect on some people. Some men never get over it. Fortunately, most adjust and put it behind them.

We stopped in a village and found an empty house to stay in for the night.

"Syng, I want the three men that refused to get involved in the fight to guard the prisoners all night. We probably should keep an eye on them. We have to know where their loyalty lies. If they won't fight, Syng, they're no good to us. We need to get rid of them."

The next morning all of the prisoners had escaped. The three soldiers I had sent to guard the prisoners were Communist sympathizers or Communist soldiers who had been sent to Japan, picked up off the street of Seoul, and were probably Communist agents.

I took their weapons away from them, told them to go join their brothers, and let them know that if I saw them again and they did not have the prisoners with them, I would shoot the lot of them.

We moved out to Pusan and arrived before noon. Lieutenant Rybolt was waiting for us. He was a little shaken.

"We're going to North Korea," he said.

I told him about the episode with the North Koreans and suggested the all men remaining be given a decoration, a Silver Star. Their actions had saved a lot of lives. He said he wanted me to go to the command post and get staff sergeant stripes.

I said, "I'll do that as soon as I find an officer to approve this citation."

"The citation is not important at this time," he said. "What is important is your getting these stripes and putting together the supplies necessary for the northern invasion."

I said, "Lieutenant, does that mean I have to give up my truck and troops and spend more time in supervision?"

"That's right, Abbott."

"We don't have one experienced man capable of setting up the switchboards," he said.

"I never know if all the lines are in working order or not. Sergeant Cutting was made communications chief when Captain Snodgrass got mad at me back in Japan. Now equipping these men, Lieutenant, is Sergeant Cutting's job."

"But, Abbott, I need you to make sure that all the supplies are in and in good working order. The other men, Sergeant Cutting included, have no idea whether the equipment's working. They know it's there, but they don't know how to make sure it's working properly. We need our inventory brought up to snuff, and we need the men to have clothing. "

"Lieutenant, I don't want to give up my truck. I'll give up Greer."

"Abbott, Greer doesn't know a damned thing."

"I know that. But he has sense enough to ask questions, and I'll help him. He'll work hard, and he's loyal, Lieutenant. I've made it my business to see that communications to the companies and outposts have been operating, and I'll continue to do that. You must make Master Sergeant Cutting do what you want done. I apologize for causing problems with Captain Snodgrass and not taking the stripes. I apologize for not being able to help you more. But you have a communications chief who is capable—he's just lazy and not a self-starter. He'll do fine. All you have to do is put the pressure on him and tell him you want the damned thing done. He won't venture out on his own. You're going to have to be assertive, Lieutenant. You have to take the bull by the horns. Your radio chief resents my suggestions. I hope Greer will provide some stability. I'll outline his job responsibilities as a wire chief and help him to achieve those goals."

"It's sad we didn't sit down and put the platoon together back in Japan. I came back to the company from the stockade to help because I felt indebted to you and Lieutenant Miley. Lieutenant Miley was gone, organized training was non-existent, and you had all new people to train. When I saw what had happened, I tried to help. Many of those I trained resented my telling them what to do or how to do it. After all, I'd been reduced in rank and now they outranked me. Cutting and Maynard took me as a threat to their potential promotions. The regimental commander sent word to me that it would be to my advantage to transfer to the regiment, but I felt obligated to come back here. It seems like I always make the wrong decisions and screw things up, Lieutenant. My transferring back into regiment would have made it possible for me to open up the communications school, and perhaps we wouldn't have as many untrained people as we've got today. But that's hindsight, Lieutenant. It's hard for me to realize we don't have one lineman capable of climbing poles. That training is desperately needed, and we will see it one day. Don't blame yourself, Lieutenant, because it was the upper command who dropped the ball on this thing. It was not your responsibility. But now we've got to play the cards we were dealt. I've always maintained that communications is the lifeblood of an Army unit, yet it's taken for granted. When someone picks up a telephone, they expect it to work. The person who made it work is never asked about. If it fails to work, then panic sets in. Often our perimeters are so spread out so that runners are impractical, and too often communications platoons are used as dumping grounds...but that's no reflection on you, Lieutenant."

"Abbott, promise me: if anything happens to me, will you take over the platoon for me?"

"Nothing is going to happen to you, Lieutenant. That's a moot question, but yes, I would be honored to do that. Thank you for asking."

“The other officers have a great deal of confidence in you, Abbott. I know the men will follow you.”

He gave Greer the stripes. I kept the Russian truck, gave up staff sergeant pay, and proceeded to get my little crew ready for North Korea and the occupation along the Yalu River. I was certain the war would not last another month and I would be out of the Army and on my way home.

The ships were in Pusan Harbor waiting to be loaded for the invasion of North Korea. We loaded the Russian truck onto the ship. We gave her a new paint job and a new white star. The ship’s painter furnished and mixed the paint, and François and Greer painted the truck. Then François painted a new star on it, a combination of our star and the Star of David.

While onboard the ship we managed to empty the shelves of the ship’s store, which made us well-equipped when we went ashore at Iowan, North Korea.

While on the ship one of the naval officers asked if I was Jewish.

I said “No. Why?”

“You have a Jewish star on your truck.”

"I'll tell you a secret: the fellow who painted that star doesn't know the difference between a Jew and a hard-shell southern Baptist."

Chapter 5
Iowon to the Fusan Reservoir

It was late October to early November 1950 at Iowan, North Korea. The 3rd Battalion headed north to the Fusan Reservoir. We had secured the beachhead without firing a shot.

Fusan was close to the Chinese border and there were rumors the Chinese would react.

I listened to Radio Moscow, which played music twenty-four hours a day, interspersed with propaganda messages. The format was the same as it had been in 1948. It was simply a different place, a different time, and different players.

In South Korea in 1948 it was Kim Il-Sung blasting the airwaves, interrupting the music: "We're going to liberate our fatherland and reunite the Korean people. We will crush the Syngman Rhee government—the puppet of the imperialist America."

Now it was the great Mao Tse-Tung vowing to "crush the American imperialists if they violate Chinese borders." The Americans had committed an act of aggression against their neighbors, killing women and children. That would not go unnoticed.

There was a new message as well: "America would not tolerate an invasion force upon its borders, such as Cuba, without a quick response. The Americans should not expect the Chinese people to tolerate a violation of its borders without a quick response. The Chinese people continue to seek a peaceful solution."

This was the same speech I had heard in South Korea in 1948. A little different tone, perhaps, but the message had not changed: "We will crush the imperialists if they threaten our borders."

When we arrived at Fusan Reservoir and met Chinese resistance we suffered some casualties. While our wounded were being evacuated by ambulance to the aid station, I was about a quarter of a mile or so away, lying in the Russian truck listening to Radio Moscow, when I heard small-arms fire coming from the valley.

There were two or three ambulances speeding down the valley. Lieutenant Rybolt, the battalion communications officer, requested my position over the radio. He advised me they were under attack and needed me, on the double. I could tell by Lieutenant Rybolt's voice he was a little shaken by it all.

"Lieutenant," I said, "It sounds to me like you have a serious situation. If that keeps up, there will probably be more wounded soldiers. Should I stay here to keep the road open so the wounded can be evacuated?"

"Abbott, get your ass up here."

"Let us understand each other, Lieutenant. You are under attack, Sir, by the Chinese forces. It is your intention, Lieutenant, to place me in harm's way. Well, in my judgment, Sir, you ought to be ashamed of yourself."

"Damn it, Abbott, I need you."

By this time, I was within a hundred feet of Lieutenant Rybolt. Small arms fire was ripping through the air and coming from all directions. Lieutenant Rybolt waved his hand for me to come closer to his position.

Within five to ten minutes, the Chinese had retreated over the hill. One of the radio operators was pinned down and we started up the hill to rescue him.

Lieutenant Rybolt was about five feet away as we approached the hill. I noticed every man alive was bunched up in one group.

"Lieutenant," I said, "One burst of machine gun fire will take every one of us out. Don't you think you ought to scatter your men just a little bit?"

"Scatter out," he said, "You heard what he said."

I sent six men to the south side, going up the hill, and six men to the north slope. The rest of the men were scattered across the hill. Any attack by the Chinese would be met with crossfire. We took the hill without resistance and secured the radio operator, who had actually shot himself in the foot.

I said, "Lieutenant, how in the world can anyone isolate a radio operator like that, leaving him without support? Who in the world made that decision?"

Lieutenant Rybolt did not respond.

A heavy snow fell that evening. Recon assured us the Chinese had left the area. Lieutenant Rybolt called me to battalion headquarters and instructed me to make sure the outpost had telephone communications.

We looked the outpost up on the map. There was a large tree at the fork in the road. Turn left at the tree three, go three hundred yards up the hill and at the top we would find the outpost. The problem was that we had a foot of new snow with a drift about three feet deep.

My Russian truck got stuck before we went a hundred yards. It was getting late. There was no time to mess with my truck. For now it would have to stay stuck in the snowdrift.

About fifty yards away was a Korean man with an ox and sled in the valley. I asked my ROK sergeant, Syngman Rhee, to call him over. I told the man I was going to use his ox and sled, and I would take good care of them and return them to him in the morning.

He protested.

I told him it was simple: I would take the ox with his permission and bring them back safe and sound, or I would take the ox and, when I was finished with them, I would shoot the ox, and if he continued to give me a problem, I would shoot him too. He was glad to let us use his ox and sled.

The men loaded the equipment on the sled and we moved over the hill. By the time we reached the top of the hill it was dark and visibility was limited.

New snow covered the road and there were signs of fresh footprints in the snow, assuring me I was on the right track. As we came to an open area I could see a small light, like a candle, about fifty yards ahead.

I had noticed movement on each side of the road for the past hundred feet or so. I was leading the ox when my Korean sergeant tapped me on the shoulder. He said, "These are Chinese soldiers."

We were within fifty feet of a large tent and directly in front of two small tents. No one spoke a word as I turned the ox around and headed back down the way we came. As we worked our way through the trees, we were more vigilant. There were Chinese troops on both sides of the road, less than fifty feet away. They were eating chow. The fact that an ox and sled were on the hill didn't bother them at all. The Chinese used oxen and sleds for their supplies. No one questioned us and we walked out scot-free.

When I got back to the battalion command post, all the battalion officers were hunched over a map.

"Gentlemen, I have a serious question. What excuse do we have for intelligence around here? You gentlemen sent me to establish telephone communications in one of our outposts, when in actuality, judging from the size of the tent, it's a Chinese battalion headquarters."

"How did you get up there with all of that equipment?"

"An ox and a sled."

They looked at each other and laughed.

I said, "Gentlemen, it might make a little more sense to double the guards in case the Chinese decide to attack." I thanked God I was not captured or killed. I was grateful for not losing my men.

I ran into Colonel McClain on my way to return the ox and sled to the old Korean man. Colonel McClain wanted to know what I was doing with the ox and sled. I told him I had borrowed them from a Korean neighbor to carry supplies to establish communications to an outpost that turned out to be a Chinese battalion headquarters.

Colonel McClain was all soldier. He listened to every word I said and then asked, "Did you see any artillery there?"

"No, Sir, all I saw was Chinese soldiers eating chow."

He turned to walk away. Then he stopped and turned around and said, "Abbott, you led that ox right up to the command post?

"Yes, Sir."

He shook his head. "Where do we get men like you, Abbott?"

I said, "Off the farm."

That story made the old man's day, and as he turned back toward his Jeep I said, "Colonel, I mean no disrespect, Sir. I'm very proud and honored to have you as a commander, and I would follow you through hell and half of Georgia. But Colonel, as big a target as you make, standing up in the Jeep in the line of fire is the dumbest thing I have ever seen a grown man do. And I've been to several county fairs. Would you please think about us shitheads? We only have one regimental commander and, personally, Colonel, I would like to see you stay alive long enough to get us out of this mess. No offense, Colonel."

"None taken, Abbott."

He went back to his Jeep and drove off. Colonel McClain and I had known each other back in Japan.

As we pulled back from Fusan, there were several Korean civilians who wanted to retreat with us. I decided not to allow them to go with us.

One of them was laughing boastfully and assured us the Chinese would be returning shortly. I told him if one Chinese soldier showed up before we left, I would shoot him and then I would shoot the Chinese. He was no longer laughing. My remarks were not quite as entertaining to him as they had been previously.

There were several women in the village who wanted to go with us as we moved south. One of our South Korean soldiers, named Chung, was preoccupied with one of them and was ignoring my orders. I was telling the men to get on the truck so we could move out before dark. He insisted on staying. A wind chime was hanging on a small piece of straw rope about a foot above his head. I shot and cut the rope holding the chime and it fell, hitting him on the head. He had a helmet on so it didn't hurt him, but it scared him. He dove face-first into the hard ground. He forgot about the Korean girl. He jumped up and ran to Sergeant Syngman Rhee and said that I had tried to kill him.

I said, "Syng, tell Chung that if I had meant to kill him, he would be dead. I had entertained the idea of removing one of his ears, but I gave that up for the wind chime."

Chung was an outstanding soldier. On the second day at the Chosin he had been shot through the mouth. He came to me to show me his wound. I hugged him and told him his mouth was too big, but that I was proud of him and loved him like a brother. He seemed very happy. He would later help me burn the trucks on the way out of the Chosin.

At Fusan the Koreans wanted me to give some civilian women a ride back to Iowan. I couldn't do that.

On our way out of the valley I went into a curve a little bit too fast and slid off the road, turning the truck over on its side, spilling wire and equipment all over the ground.

Suddenly there was screaming coming from under the reels of wire. Two of our South Korean soldiers were trapped. We frantically started to remove the wire. When we got down to them and removed the wire, we found them trapped between the reels of wire. They had not been touched.

There was a telephone pole not twenty yards away. I climbed the pole, tapped the wire, and got straight into the motor pool. When I asked for a wrecker to come pick up the truck, the voice came back, "Is that your Russian truck, Abbott? We're not going to send a wrecker out to get that truck running!"

“Pass me through to the command post and let me speak to Major Storm.”

I said, “Major, we have a problem. The motor pool doesn’t want to bother with this old truck. That's understandable, Major, but what do we do with all this equipment?”

“Help will be there within the hour, Abbott.”

Within twenty-five minutes we were up and running again.

The following day we stopped for lunch near a beautiful little creek, built a fire, and were heating our rations. As I looked out over the valley I saw Major Storm and a new captain walking toward us. I walked out to greet them.

The captain wanted to know what we were doing out of the convoy. I told him we needed to fix a tire, and while we were stopped we would have lunch.

“You gentlemen are welcome to join us. We have beans—Harry Truman's finest beans.”

The captain, evidently trying to make an impression on the major, jumped on my case about not helping the Koreans fix the tire.

"Captain, the day and hour I have to do these Koreans' work, you can find somebody else to baby-sit them." I saluted and walked away.

Major Storm said, "Captain, you'd better leave that little son of a bitch alone. You're going to need him."

We finished eating lunch and moved out as the two officers watched us leave.

I, of course, did not need to baby-sit my Korean troops. Any South Korean troops who had proved unworthy in battle in South Korea when the battalion was moving south to Pusan were kicked out of the platoon and removed from duty.

We moved out across the mountain. The road was steep and narrow. At one point I placed all the men on the left side of the truck. If we were to start sliding over the mountain, the fall would be about five hundred feet straight down. This way the men could simply jump off the truck. Further on we had to go around a big rock, and at one point we had two wheels on thin air, but we made it over the mountain and into the 31st Infantry compound.

I had just gotten into my sleeping bag when Lieutenant Rybolt came looking for me. He had a Combat Infantry Badge for me.

"Lieutenant," I said, "pin it on the sleeping bag and let's go to sleep."

He stepped out of the tent and threw the badge away, calling me an ungrateful bastard.

I said, "Goodnight, Lieutenant. I love you too."

Greer, Abbott, and "Chief"
Photo courtesy of the family of Joe Mack Harper

Chapter 6
To the Chosin Reservoir

Early the next morning, November 27th, the 3rd Battalion headed north to the eastern side of the Chosin Reservoir, some sixty miles away.

I had my usual Korean soldiers and the old sergeant, my trusted good friend whom I respectfully called Sergeant Syngman Rhee, after the South Korean president. He was an outstanding soldier, a man who saved my life on three different occasions.

I had passed up promotion to staff sergeant. I told Lieutenant Rybolt to give Greer the stripes and I would teach him how to be a wire chief. Staff Sergeant Greer, the wire chief, was a good man but knew absolutely nothing about communications. I was happy with my Russian truck and my lot. I had a crew of good men and was self-sufficient. However, that day was a sad day for me. My gut told me we were heading for deep, deep trouble. I had developed that burning in my gut at the Fusan Reservoir earlier that month and for some reason, I knew the Chinese were here to stay. We had not seen the last of the good Chinese.

Sergeant Greer had allowed a medic to ride in our truck. He knew only wiremen were to ride in the communications truck. He did not understand that the medic was supposed to be with the line troops where he would be needed. Our trained communications people were ready to go to work once we hit our staging area and the bivouac area.

I realized what a mistake I had made. I had done Lieutenant Rybolt a terrible injustice by not taking the stripes. Greer simply did not know any better. Furthermore, it was unfair to Sergeant Greer. He knew he was in over his head, so he would not let me out of his sight. I was doing a lot of soul-searching on this drive.

I needed a hibachi. It was cold. I pulled off the road in front of a group of farmhouses and ran over to get a hibachi from one of the huts. Unfortunately, the hibachi was hot and I had no gloves so I had to leave it there.

The hair on the back of my neck was standing on end. I could picture in my mind the Chinese swarming down on us. I remembered the carefree attitude among the officers at Fusan. The attack had been small, but I knew there were thousands of

Chinese that did not attack. I did not know what was in store for us today, and it bothered me.

When the convoy stopped I told Syng I had a funny feeling, and told him to make sure our weapons and ammunition were where we could get our hands on them. We normally stored our weapons under reels of wire and covered them with a tarp to keep them out of the weather. I wanted to make sure we had ammunition for our rocket launcher.

"Syng," I said, "For heaven's sake, don't say anything to anyone. They already think I'm crazy."

I had made sure that all of our people had extra blankets, sleeping bags, and extra rations, as well as cans of fresh water. I also had made sure all of our men had their winter clothing and boot packs.

One little Korean's feet were too small for boot packs so he had to wear boots. The boots were too big for him but we made sure he had extra socks.

We found four extra fur caps which meant we were in good shape. We had stocked up on gasoline and oil. We had a fifty-gallon barrel of extra gasoline, and I was able to secure two extra tarps from the ship's storage on the way to the North Korean landing.

Somehow that uneasy feeling kept nagging at me. I pulled out of the convoy and stopped for lunch. I don't believe we fed the medic that day. We needed to kill some time and the convoy was about ten miles ahead of us. The main convoy had gone through the Marines' compound.

I asked a Marine Corps captain where the 31st Infantry was and he said it was over the mountain, at the inlet of the reservoir.

"Captain," I said, "It looks to me like you expect serious problems, the way your men are digging in." He looked at me with a very serious look.

"Son, we expect a full-press Chinese attack tonight. Your people will be right in the thick of it up there at the inlet."

"Is it safe to go over the mountain tonight, Captain?"

"I don't know. But if it were me, I would hold off until morning."

As I walked back to the truck, I was thinking about a safe place for us to spend the night.

I told Syng it might be better for us to stay until morning and then cross the mountain. Then I saw the medic and realized there was no way we could stay for the night. I reminded Greer what a damned fool he'd been to allow that medic to come with us. Now we had to cross the mountain because that medic would be needed if there were an attack.

It was late dusk as we went over the mountain. The road was steep and crooked; it climbed rapidly. We came to a hairpin curve near the top of the mountain. I stopped, got out, and looked over the valley. Syng came to me and said, “Chinese.” I looked at the hill to my right. It was crawling with Chinese. We headed over the mountain.

We arrived at the 3rd Battalion headquarters. Nothing had been done, not even the switchboard was completely set up. The company command posts were without telephone communications and the 57th Field Artillery had no communications. Sergeant Cutting should have had communications established for those command posts before dark.

We had to rectify the situation, and my men worked until midnight. We also recruited other men from within the company to expedite the installation. I told Greer to have someone run a line to the 57th Field Artillery and to check all switchboard lines. Everything was complete, except for one line to the anti-aircraft battery.

I was angry, tired, and disgusted with the whole damned bunch.

One of the officers, a captain, said, "The Chinese will cut the wire tonight. We need it overhead, Abbott." There were poles already in place.

I was the only man who could climb the poles. I told the captain I didn't have time to do it and if he could figure out a way to get the wire up on that pole, to have at it.

Another officer overheard the conversation between the captain and me, and I apologized for the inability of the communications platoon to do its job. But time was running out. He said, "It doesn't make any difference. We're jumping off early in the morning."

I went back to the command post.

Lieutenant Rybolt told me the same thing, and I said, "Lieutenant, do those Chinese on that hill know that we are attacking in the morning?"

"In no uncertain terms."

I was assured there were no Chinese in the area. There had been reconnaissance patrols out all day and they had seen no sign of Chinese.

I said, "That's really sad, Lieutenant. They don't want you to know they're there. Will you tell the colonel, or shall I, that the Chinese are combing those mountains?"

Colonel Reilly must not have been notified. At any rate, he did not tighten up his defense and his perimeters, nor did he put the battalion on alert.

It was almost midnight when I crawled into my sleeping bag. There was a fairly deep drainage ditch alongside the communications hut. Since I had no foxhole, I lay down in the ditch and went to sleep.

I was awakened by bullets hitting the ground all around me. I got out of the sleeping bag. The bugles began to blow and small arms fire began coming in from all directions. Our fifty-caliber automatic weapons had begun to respond. I crawled over the bank and saw movement about fifty feet away. There were five or six Chinese gathered next to a truck.

I threw two grenades; they went off, and there were some Chinese crawling off, and others screaming who were promptly shot. Machine-gun fire began to pick up dirt around my feet.

I jumped down the bank and someone said, "Get back up there!"

I turned around and saw people running into the command post area.

The sergeant who had said, “Get back up there!” saw who I was and then said, “Lord God, Abbott! I've lost all my men, my mortars, and I don't know what happened to them.”

“Sergeant,” I said, “your men are probably scattered someplace. They probably found a foxhole to hide in. We’ll find them in the morning, and we’ll get your weapons back because we need those mortars. In the meantime, Sergeant, get yourself together, find a rifle, and help us.”

He said, “Okay.”

I never saw him again.

I went around to the north side of the building. The Chinese were trying to climb up a steep hill to the command post. It was like shooting fish in a barrel. They just kept coming. I used my last hand grenades. They were running back down the hill. My M-1 rifle stopped firing and I went back to the communications hut in search of a rifle.

The Chinese pulled back at daybreak. We had done a fair job of holding them off. The thought went through my mind: “How were they able to break through our lines so quickly?” The only thing

I could think of was that the whole damned bunch, from the battalion commander all the way down, had believed that bull of attacking in the morning. Everyone was asleep and the Chinese had simply walked in on us. How could this happen? What was so important that an infantry unit had to be rushed into an ambush?

The Marines that were on the other side of the mountain were ready for the attack. We were not. We didn't have time to get set up. Our losses were heavy.

I saw a man whose legs had been mutilated by a fifty-caliber machine gun. He continued firing his rifle until life left his body. I wondered why a man would continue to fire after he had been so severely injured. His words continued to haunt me: "You son of a bitch! You son of a bitch!" That man's name, I believe, was Harper.

There were a lot of good men who died at the Chosin. They have no markers and they have no flag-draped caskets. When spring came at the Chosin, their bodies just rotted away. Without them, no one would have walked out of there alive. But that's war. You win some and you lose some.

Nothing makes a man live longer in combat than a cool head. You have only one job, and that is to kill the enemy. It bothered me to see men standing around talking in an open field surrounded by enemy snipers. If the Chinese had been better marksmen, a lot more men would have lost their lives.

Everyone seemed to be looking for someone or something. They ended up talking to each other in the middle of the field. We were lucky that day. The Chinese were confused and disorganized. They suffered heavy losses and retreated into the hills very rapidly.

From my vantage point that night, halfway up the hill, at the 3rd Battalion headquarters, I could see the battle unfold.

There were places on the north side of the compound where the hill dropped out of range of our artillery and automatic fire, and they made good points of attack for the Chinese. Attacking from the north required the Chinese to climb up a steep hill, which proved costly. The north side provided cover from our automatic fire, but it provided no cover from small arms fire, and our men took advantage of the exposure. We fired directly down on the Chinese. The Chinese paid a heavy price.

I used my last four grenades the first night. There was no central location to get ammunition. It was a sad situation. Had our men had a full supply of grenades, we could have brought the wrath of hell down on the Chinese troops as they tried to enter the command post from the north side and up that steep hill down there in the valley.

As I looked down in the valley, I could see the Chinese overrunning the 57th Field Artillery. Once the Chinese had passed their position, the men who had hidden under the guns as the Chinese ran past them mounted their guns and fired point-blank into the rear of the enemy. I was so proud of the artillerymen as they defended their position.

The command post area had received heavy mortar fire, and the wounded included the commanding officer. The mortar round had wounded Lieutenant Colonel Reilly pretty heavily with shrapnel. The concussion seemed to cause him a great deal of problems and he was in bad shape. He wasn't quite dead, but he was in serious condition. He was removed from the compound.

I found the carbine I needed and headed back to the north side of the compound.

The bugles started blowing and the Chinese started retreating from the north slope rapidly. Lieutenant Traywick was calling my name. Before I could get back to where he was, he was moving out in a counterattack. I could see the Chinese stragglers in full retreat.

I did not join the fight. Lieutenant Traywick, upon his return, asked where I was. I said, "Protecting my ass."

He said, "I needed you, you know."

"I'm sorry," I said, "Y'all did just fine without me. Perhaps next time, Lieutenant." He went into the command post. I felt badly that I had disappointed him. Lieutenant Traywick was one of the best officers in the battalion. Any way I could help him, I was very happy to do so.

After daybreak on the second day, men and officers alike were walking around, bunching up and discussing the previous night's battle without consideration for sniper fire. By the second day, the Chinese had established firing positions on the high ground, and snipers were experiencing considerable success. Several men and some of the officers were hit by sniper fire.

I walked across the compound and noticed a master sergeant sitting on a step, directly exposed to fire. He had done well the night before, fighting the Chinese. The Chinese had overtaken his foxhole and he had killed twelve Chinese with a .45 pistol. I congratulated him for his actions and thanked him.

He said, "It's a good weapon, Abbott."

I told the sergeant he needed to get off that step. He said he thought he was safe there, even though I told him the Chinese had zeroed in on him. Sometime later I walked back by and I could see the pain in the sergeant's face. His words were, "Why would they want to shoot me, Abbott?"

"Sergeant, they shot you because you're available. You invited that bullet, Sergeant. You spent a lot of time in World War II. Did you not learn that a sniper would always choose a stationary target as opposed to one that's moving? You have been sitting there, probably, for the last three or four hours, and if you don't move out of the line of fire they'll finish the job."

I offered to help him to the aid station. He refused. I don't know what happened to him. He moved off the step and I never saw him again.

Our perimeter had tightened up to where it should have been in the first place.

Lieutenant Rybolt had been hit. A large portion of his forehead had been blown off, but he was still conscious in sickbay. The men needed help desperately.

"I'll have to take Rybolt's place and do what I can to help these men," I thought.

The Chinese snipers had established their target areas where the sergeant had gotten hit. The far north corner, at the edge of the compound, was receiving fire at any movement. I had suggested that all men simply avoid standing in these areas for any length of time. "Move past these areas rapidly," I cautioned. And I told the men not to bunch up at the wall, but apparently they did not heed my warning.

Another attack came the second night. It was a little earlier than the night before. We were in a little better position to defend the attack than before. Everyone had pulled back and established a good defensive position. Without realizing it, I had actually taken over Lieutenant Rybolt's position as a platoon leader, but had not requested it be made official. Some of the men resented my giving them orders, as they all outranked me.

We were under heavy fire and I had gone around to the southern corner of the wall at the edge of the compound. Every movement got an immediate enemy fire response. I was heading back to the north side of the compound when I ran into Sergeant Maynard, our radio chief. He evidently resented my giving orders to the men. When I told him to stay away from the south end of the wall, he said I had no authority to tell him what to do.

"Well, Maynard, do as you damned please. But if you go back around there, I just want to tell you, it's been nice knowing you for these past three years."

It was less than five minutes later when I was told Sergeant Maynard got it right between the eyes and died instantly. He would not listen.

"Damn it!" I said.

Maynard had been a private when I was communications chief back at Crawford. I had made him the radio chief and now he outranked me and resented my telling him what to do. It cost him his life and made me feel very, very bad. I thought, "What a waste. Had I made him angry enough that he just wanted to prove a point?"

About an hour had passed. The north side of the compound was coming under heavy attack. I ran over to that side. After two or three rounds, my M-1 rifle I had decided to freeze up. It would not eject the shells. I started looking for another rifle, went into the communications shack where the switchboard was located, and found Sergeant Cutting, the communications chief, sitting in the corner grieving over the loss of his good friend, Sergeant Maynard.

I said, "Cutting, I'm sorry about Maynard. But you know we can't bring him back." I sat down against the wall just as a mortar shell landed next to the wall, and it sending me flying into the air. I landed hard on my back. I thought my neck was surely broken.

I rubbed the back of my neck and got a handful of blood and dirt. The same was true of my face and head. I had forgotten the reason I came into the communications shack to begin with.

The next thing I remember was Gus, the man operating the switchboard, pleading with me not to send him outside. I suddenly remembered why I had come into the communications shack. I assured Gus he was just where I needed him to be. I remember his words: “Lord, God, Abbott...don't send me out there in the line of fire.”

“Gus,” I said, “just keep your head down. I need you here. Gus, do you have a rifle?” He said, “No, I don't have a rifle.”

I asked Sergeant Cutting if he had a rifle, and he said no. I looked at Cutting. He was in no condition to do anything but stay where he was. “Cutting, you stay here and get your shit together. Everything will be okay. I've got to get back out there and see what I can do.”

I walked outside and found the battalion commander, Colonel Reilly, incoherent and repeating to himself, “I want to stay with my men, I want to stay with my men...”

I crawled over to him and said, “Colonel, I don't want to hear anymore of that bullshit. You’ll never know how proud I am of you, but it would be such a waste for you to stay here and try to command with your feet filled with shrapnel and unable to walk. The important thing now, Colonel, is that you live to command again.”

He said, “Abbott, if we get out of here alive, I'm going to give you a Silver Star.”

"That's very kind of you, Sir, but getting out of here is a major problem and a decoration is the furthest thing from my mind at this point in time. At any rate, there are those much more deserving than I, Sir. I'll do all I can to help keep things moving."

The next day a helicopter landed to evacuate the colonel. He was under orders from General Barr, the 7th Infantry Division commander, to leave. Sergeant Cope asked if I would like to help carry the colonel to the helicopter, even though he had enough men to do it.

"Yes," I said, "Do you need another man?"

Then someone else volunteered and I said, "Let him do it."

Then I said, "Good luck, Colonel."

They loaded Colonel Reilly on the helicopter and Major Storm took over command of the battalion. Lieutenant Rybolt was the only officer still alive whose wound was life-threatening. I felt he should have been evacuated with the colonel.

I felt then and I feel now that Colonel Reilly should have turned over his command to Major Storm when he was hit the first time. Major Storm would have had more time to have gotten his men together and put his command together for the breakout. Instead the breakout was a last-minute decision, based upon our inability to hold another night under Chinese attack. It was a Colonel Faith's decision, I'm told.

We were out of ammunition. Many of the M-1 rifles would not fire because of the cold weather. We had an airdrop of machine guns earlier with no ammunition to go with them.

We also got an airdrop of toilet paper. The only black soldier at the Chosin was part of that airdrop. He had gotten tangled up in the riggings and the ropes holding the toilet paper, and they pulled him off the plane. He landed safely and was captured by the Chinese, and later killed by friendly fire.

I had a machine gun without ammunition. There's nothing more demoralizing to a soldier in combat than to be low on or out of ammunition. A soldier's lifeline is his weapon; his ability to kill the enemy and to stay alive depends on it.

Just before daybreak on the third day of the battle, I felt I had lost track of time. One of the Korean troops came to me wearing a Chinese jacket and said, "You can do no more here, Abbott. We leave now."

"No," I said, "I cannot desert the wounded. We must try to help them."

That was the first indication we were going to break out. I told Syng to go help all he could. He said it was too late for help now. I wished him well, and he left.

I moved to the left side of the building, watching Chinese troops march three abreast down the road toward the inlet.

We were totally helpless. We had no way of interrupting the troops. A machine gun was lying there within three feet of us, but we were without ammunition to fire it. We received sniper fire and two bullets went over our heads very closely. We hugged the ground. Pryor punched me in the side and yelled, "Are you hit? Are you hit?"

"No, Pryor, I'm just saying a word of prayer for you."

About that time, a lieutenant called my name. Pryor said, "Someone's calling you."

I said "Yes, and I better go see what he wants."

The lieutenant had spotted a forward observer high up on the north hill and wanted to know if I could hit him. I said, "Not with this carbine."

I began to look for an M-1 rifle. I found one, but it was the one I had discarded the night before because it would not fire. I saw the O3 rifles I had captured in South Korea when we were involved in a firefight there. I knew it would shoot straight and it had a clip of ammunition.

At the crack of the rifle the forward observer stood up straight, and then rolled down the hill. The lieutenant picked me up, put me on his shoulders, and carried me around the compound saying, "You got him, Son! You got him, Son!"

"Lieutenant," I said, "For God's sake, put me down! We have a lot of work to do." He told me he thought we could get out of there. He had been trapped in World War II and gotten out of it okay.

"This is not World War II, Lieutenant. If we don't get out, let's make sure the enemy will remember we were here. Let's make sure they don't forget that." By this time the wounded had all been loaded and were moving out.

I looked around and I saw my old friend, Syngman Rhee. He was here, carrying a full Chinese uniform this time, telling me to put it on and leave, as I could do no more.

I said, " Syng, we will go, as soon as I burn the trucks." We could not allow the Chinese to use our vehicles to attack us from the rear.

The wounded had been loaded onto trucks. At that point in time, I was not aware that the bridge at the inlet had been blown up.

Corporal Keyes asked me to cover him while he replaced a broken telephone line to the company command post. I told him no. He had a reel of wire that clicked with every revolution and the click was very loud. We were taking sniper fire from all directions. I told him he would not get twenty feet before getting shot. Keyes told the company commander I would not cover him.

The company commander said, "He belongs to Major Storm."

Major Storm said, "What in the hell's wrong, Abbott?"

I said, "Nothing's wrong, Major. I'll take care of it. Keyes, give me the wire, put your jacket over the reel, and keep it quiet so we don't draw attention. I'll take the wire to the command post."

I made it to within twenty feet of the foxhole that contained the command post. Two or three bullets sang past my head, and two or three hit the ground. I dove for cover. I did not move a muscle I just laid there. Suddenly, all firing stopped. The snipers were secure in the thought they had killed me, and I took the opportunity to dash into the command post. There was only one officer, a first lieutenant. He had been shot through both arms. He said, "Oh, it's you, Abbott."

I said, "Yes, Sir. Who did you think it was, Santa Claus? It's a little early for that, Lieutenant, even for him." We laughed.

He said, "How are you doing?"

I said, "Somewhat better than you, Sir," and I pointed to his arms. "But God has smiled on us this day, Lieutenant, because at least we're still alive."

I connected the telephone and said, "I have to go now." I went back up the hill without drawing fire from the Chinese.

Corporal Keyes asked me what I wanted him to do. I told him to cut the wire from the switchboard and throw it over the hill, and then get the hell off the hill.

The switchboard was no longer in operation and the command post had a direct line to Major Storm.

I took one man, a Korean, and had him set fire to the trucks that had been left behind. Once that had been competed, I remembered Keyes. I ran back up the hill to get him. He was disconnecting the switchboard and humming a song as though he had all day to do it. I hollered, "Keyes, leave everything, bring your rifle, and run as fast as you can!"

Keyes turned to me, totally unaware of what was going on around him. He said, "What's goin' on?"

"Just follow me, Keyes." I stepped through the door, stepped back, and was surrounded by at least twenty-five Chinese soldiers. "Stay close, Keyes. Stay close to me." As I looked to my right, I saw the commanding officer of the Chinese forces entering the compound. This was the man I had seen two days ago, and also on the second day of the battle, going to a hut about a hundred yards up the hill. I was sure he was directing the battle from up there.

I had read the tactics of the great Kahn. I could see him in my mind, sitting on that white horse, directing the battle. "This was a great man," I thought. "he's clean-shaven and his uniform is spotless." I had waited two days to get a shot at him.

I had tried to get my hands on a 3.5 rocket launcher the day before. I had found one but had no ammunition for it. Now here he was within thirty feet of me, looking down the barrel of my rifle. He stopped dead in his tracks, looking straight at me. He brought his hand up as though he signaled to his men to stay back. I had a burning desire to kill this man, but now, face to face with him, I had nothing but admiration for him for some reason. I came to attention, saluted, bowed, and said, "My congratulations to you, Sir, you have done a good job. The job has been outstanding. If I had a medal and the authority to do so, I would decorate you."

He bowed, "I do wish you well, Sir."

I would be less than candid if I did not admit that his being surrounded by so many troops had influenced whether or not I would actually pull the trigger.

I turned and pushed two Chinese soldiers aside. He gave an order, evidently telling them to step back and let me pass, and they stepped aside. I walked out untouched. After leaving the compound as I headed down the hill, I could see Chinese soldiers coming from all directions. "Run, Keyes," I said. I must have broken all speed records that afternoon.

The Chinese were now on my left and right flanks and were firing heavily. I heard Keyes holler, "Oh!" and he fell right behind me.

When I got to the bottom of the hill, I ran and jumped into a foxhole. I found a young soldier sitting in the corner of the foxhole.

He said, "My God...they can't kill you!" He had evidently watched the situation unfold as Keyes and I came off the hill. He continued, "If one man gets out of here alive, it will be you, Abbott. They can't kill you."

"The hell they can't kill me!" I said. "They can and they will kill me. And you can be sure if you don't get out of this hole, a Chinese soldier will put a bayonet through your rear end. Now you get your ass out of here. Get behind that tank over there until you can get your head on straight. Then get a rifle and help us."

He got out of the hole, got behind the tank, and I never saw him again that day.

I walked around to the corner. There were several soldiers taking cover near the road. Among them was Private First Class Davis, who was shot in the gut and was dying. Staff Sergeant Greer was shot through both arms; no bones broken, just clean flesh wounds. I asked Greer if he could move his hands and raise his arms. He said he could.

"Abbott," he said, "We came in together, let's go out together."

"Greer," I said, "You're not hurt that bad. Find a rifle and help us. Davis, you've been shot in the gut. That's a very serious wound. I suggest you get on one of those trucks."

"No, Abbott," he said. "I'll walk as long as I can."

Greer got out of there, but Davis died that morning. Greer had made no attempt to find a rifle or offer assistance.

I was a little disappointed in Greer. I had given up a promotion to enable him to make staff sergeant. He was promoted only after I had agreed to train him and convinced Lieutenant Rybolt he could handle the job. In all fairness, he never really got the chance to learn, and he made a good account of himself at the Chosin, even though he faltered a little bit at the end. Greer, I think, thought I had lost my mind. He assured me I could do no more to help the wounded.

I walked up the column, past all the trucks loaded with wounded men. One tank was sitting there, burning fuel, and was not engaged in the fight at all. It was wedged in between two trucks and was useless to defend our position.

I continued up the road and found about thirty men all bunched up. I stopped to ask a tech sergeant what was going on, and he got excited when he saw me. He said, "I'm sure glad to see you. You're too damned mean to die. Stay with us, Abbott."

I inquired as to the whereabouts of the company commander and the other officers. He told me they were up at the front of the column. I lay down beside the sergeant for a moment, watching the Chinese move around the hill toward our rear position. The enemy movement was a continuous flow of soldiers, as far as you could see.

I said, "Sergeant, why are we stopped?" He said the bridge was blown. "Well, hell, Sergeant, we can't stay here. We have to get the column moving. This would be like shooting fish in a barrel."

I decided to move forward to find someone to help get the column moving and to get those trucks that were loaded with the wounded around that bridge. The men were bunched up like sheep in a fence corner during a thunderstorm, scared and without direction.

As I looked to the rear, I could see a Chinese soldier with a machine gun, and another soldier about ten feet behind him as his ammunition bearer.

"Sergeant," I said, "I need to get that machine gun."

I knew if he set up in our rear, we were in deep trouble. I thought if I could get back where I could get a clear shot, I might get him before he got his gun set up. I ran back down past the column and past the last truck. He still had the machine gun on his shoulder. He saw me, threw his gun down, and jumped into a ditch for cover, leaving his machine gun. The ammunition bearer evidently did not see me. I had a good shot and I took it. He fell down face-first. The first man made no attempt to return to his gun or to pick up the ammunition. As I walked back, I said, "Someone ought to get that machine gun and the ammunition," but no one did.

I could not convince the wounded who were crawling on the ice to take cover near the road. All they would do was tell me I was drawing fire. Of course I knew I was drawing fire, but it was important that these men get off the ice and either get on trucks or take cover next to the road. Those who could walk should walk on past the convoy. It was clear to me if I could get the stopped convoy to back up and go around the bridge, we had a good shot at getting out of there.

We had a tank to pull the truck off the ice in case we got stuck.

I looked off to the left of the ridge and the hillside was covered with Chinese. By now they were crawling like flies, heading toward our rear position. The only hope we had, I thought, was to back the column up and go around the bridge. Why had there not been an attempt to get that column moving around the bridge?

The tech sergeant from Harlan County, Kentucky, who had just told me I was too mean to die, was with that group. He had told me the company commander was up ahead.

The men would just look at you as if you were crazy. I had entertained the thought of shooting one just to get attention.

Suddenly I felt like a fool and asked God to forgive me for being so stupid not to have volunteered to replace Lieutenant Rybolt when he got hit. I should have made it official and had some authority. I was sure Colonel Reilly would have granted my request for a field commission. That would have put his command in order and would have given the battalion another officer; admittedly not much of an officer, but a warm body at any rate.

These men desperately needed guidance. I thought I would go back to the tech sergeant and pass my thoughts on moving the column around the bridge, and talk with the tank commander about pulling them out. As I started back to talk to the sergeant, I heard someone say to the Lord, "God, this is old Captain Kitts. Tell them to hold their fire—I have to cross this field." And he took off like a rabbit with a bluetick coonhound on his tail.

There were three or four of us standing there. I said, "Let's go," and I called out, "Captain Kitts, where are you going? Wait for us."

He evidently did not hear me. Kitts just kept going, and by this time I was exhausted and angry, and I said, "That son of a bitch that gave the order 'every man for himself' ought to be shot. How in the hell can they expect these men to protect the rear of this column without some sort of leadership?"

Three of us were following Captain Kitts. We began to take extensive fire from our left flank. I realized we had ventured too close to the ice and had lost our cover. Just as I turned to return to the road and have cover, I was hit. The impact sent me down hard on my right shoulder, digging into a broken piece of ice. I could feel the pain in my right arm and shoulder, and my right arm went numb. My back felt like it was on fire. Air seemed to be coming in through the hole in my back, and I could barely breathe.

My head began moving backwards uncontrollably, pulling, pulling, pulling. I felt myself blacking out. I grabbed handfuls of snow and put them on my face while at the same time I asked the medic to plug the hole in my back. He wrapped an Ace bandage around my rib cage very tightly, and that allowed me to breathe with some comfort. I felt like the top of my head was trying to come off. The medic said he thought it was just a flesh wound and handed me my rifle.

I said, “Take my rifle. It has about half a clip in it. Give it to someone who can use it, but don't let the Chinese get it.”

They started to leave me and I said, “Put me on my feet.”

The medic said, “You can't walk, Abbott.”

“Help me to my feet!” I said.

The two of them helped me to my feet and I made it to one of the trucks carrying the wounded. I crawled up on the trailer and saw my radio operator,

Bressler, still standing on the running board of the truck.

He said, “Abbott, are you hit?”

I said, “Yes; hard.”

As I climbed into the bed, I said to Bressler “Get off the truck and move out before darkness catches you. The column needs help.”

The discussion with Bressler was the last thing I remember until I heard enemy gunfire coming into the truck. The sound of bullets hitting dead bodies and live people is incomparable to any other sound in the world. Men were being hit with rifle bullets from above and they were screaming with pain.

I could hear men walking around the truck, and I called out for someone to pull the two bodies off of me so I could get up and go with them. I did not have enough strength to move the bodies myself. My call for help went unanswered.

Suddenly someone climbed up on the truck and said to the man who was screaming, "I'm giving you a direct order to shut your damned mouth.” I recognized the voice as that of Captain Kitts.

I said, “Captain? Is that you, Captain Kitts?”

“Yes.”

“Are you hit?”

"Yes, I'm hit very hard. I'm hit in the chest."

I had assumed Captain Kitts had died.

When the Chinese captured the truck and dropped the tailgate, I fell out, hitting the ground. Captain Kitts was not there. The Chinese had captured several men and were marching them to the rear when they stopped and picked us up. I looked around and there was no Captain Kitts.

I had nothing covering my upper body except a small jacket draped over my left shoulder. My shirt was gone. I had not been able to get it on again after I was hit. There was blood all over me. I wiped my face with my good hand and found more blood. I looked around and saw a sleeping bag on the truck and asked Sergeant Cutting to hand it to me. I pulled it over my head, and pulled it around my chest.

Sergeant Cutting saved my life that day by handing me that sleeping bag. The Chinese officer looked at my chest and pulled my bandage back. He thought I was shot straight through, in the chest and out the back. He couldn't understand how I was still living.

One of the Chinese guards came over, pulled out a package of Lucky Strike cigarettes, handed me one, and showed me his Zippo lighter. He had captured it and was very proud of it. After about three puffs on the cigarette, I passed it back to another GI. The Chinese quickly took the cigarette back.

"Sleep."

I said, "No," and finished the cigarette.

How strange it was. I thought, "Here I am, thousands of miles from home, fighting in a war, a captive of that war, and the enemy has singled me out and given me a cigarette and denied the others the same privilege."

That Chinese soldier must have been seven feet tall. He kept telling me I was going to sleep. He would say so each time and I would say each time, "No, I'm not goin' to sleep."

We had walked about three quarters of a mile or so when we stopped in an old school building. I had lost a lot of blood. The fall from the bed of the truck and hitting the ground had caused the back wounds to start bleeding again, but it had now stopped.

Just at dawn one of our planes came in and struck the schoolhouse, hitting me again, this time in the left side of the chest. The blood gushed. "Damn," I thought, "I'm shot again." Luckily, in both cases I had been shot, it was a flesh wound. I rose up to look around and used the sleeve of my jacket to apply pressure to the wound.

I could not move my right arm and the pain in my shoulder was unbelievable. I quickly rolled over on my left side, and as I did I saw blood all over my sleeping bag, which led me to think I'd been hit somewhere else. When I rolled over and raised my head, I could see where the blood had come from. The man lying next to me had taken a direct hit. The whole top of his head had been blown off.

I fell asleep. I had not been asleep very long when I was being pushed and told to get up. I pulled myself up with the aid of the pole in the middle of the room and managed to join the others.

We were marched into the hills and formed a circle around an old bomb crater. The Chinese had set up a machine gun facing the crater. All POWs were directed to place all their personal belongings into the crater.

The obvious thought was, "This is the end of the road. We're all going to be shot this morning." Suddenly Lieutenant Connors, the chaplain, approached the Chinese and said, "Take my life and spare the lives of these brave men."

The chaplain's words had such an effect on the

Chinese that the machine gun was removed and the men were marched back to the old school building.

When we arrived at the school building, I passed out. When I awoke, everyone had gone, and the dead man's body had been removed.

I worked myself up onto my feet with the use of the pole that was supporting the building, and saw a row of millet balls sitting on the ledge of the window across the room. I worked my way over to them and found that they were frozen, with a crust of ice on the outside. The crust of ice satisfied the desire for water, and the millet in the balls satisfied the hunger. There seemed to be a substantial amount of energy in the balls.

Suddenly there was a voice coming from the far end of the building behind the feeding trough. "Is there anyone over there?"

I answered, "Yes."

It was the chaplain. The Chinese had left him behind with me, and I thought what a blessing it was to see him.

The Chinese had left me without shoes, jacket, or shirt and it was fifteen or twenty degrees below zero. They had also removed the bandage from around my body, leaving my back and chest wounds exposed to the elements.

I wondered what I was going to do. Walking in bare feet in snow at that temperature was out of the question.

The chaplain asked me what I planned to do to get out of there. I advised him that I was going to walk. I gave the chaplain one of the millet balls and told him they were good to eat. He seemed to enjoy the millet.

"So what do we do now?"

Chaplain," I said, "I'm going to get you out of here."

"Just how do you plan to do that, Abbott? You have no shoes; you'll lose your feet before you get a mile."

"I don't need a mile, Chaplain, just about two hundred yards. If I can walk two hundred yards down the road, there are about twelve dead GIs. I can get shoes and clothing from them."

The chaplain seemed to resent removing clothing from the dead. "Chaplain, you, of all people, know that God works in mysterious ways. Perhaps it was in God's plan that those dead soldiers provide us with clothing so we might survive the cold weather and make it back home."

He seemed to accept my logic.

"Now, Abbott; what do we do now?"

"First we must get something to wrap my feet so I might make it to those soldiers down there. I know it's a hell of a thing to ask you, Chaplain, but if we could get the sleeves out of your shirt, I could stuff straw in them and make me a pair of shoes."

The chaplain agreed. We went to work on removing one of the sleeves from his shirt. Getting that sleeve off turned out to be a very difficult job. Our dog tags worked fairly well. We cut a hole in the sleeve, and once that was accomplished it came off quite easily. I tied one end of the sleeve, put my foot in it and packed it full of straw.

"Chaplain," I said. "When I get out of here, I'll build a frame and hang these shoes in my living room for my children and grandchildren to see." He laughed.

Just as we completed one shoe, the Chinese walked into the building. The same Chinese that gave me the cigarette when I was first captured took us out and tied us on a sled. We rode all night.

The sled was driving up a frozen creek bed with large rocks and it did not have a solid bottom. The bottom was made of bamboo sticks tied about a foot apart with grass ropes. The poles kept sliding down and we kept sliding down in between the poles.

The chaplain was thrown off the sled when we hit a large rock, and was being dragged behind the sled until the Chinese stopped and reloaded him. This time the Chinese tied him down very, very well. The dragging through the snow and ice may have been too much for the old chaplain. It may have contributed heavily to his death.

We stopped just at daybreak in front of one of the buildings were the POWs were being held. This camp later became known as Death Valley.

I inquired as to how he was doing as I had done so many times before throughout the night, and his reply was always, "God is with me, my son." This time the chaplain did not answer for at least a minute. When he did finally respond, he said, "Promise me, Abbott, that you will go visit my mother and tell her that I did not kill anybody. That will mean a great deal to her, Abbott."

I said, "Chaplain, you can tell her yourself when the war's over."

He said, "Abbott, I'm going to go be with the Lord. I'm frozen to death." And he just stopped breathing.

The Chinese rolled the chaplain off the sled. I crawled around the sled, took his head in my arms and held him, and said a prayer: "God, you have called one of yours home this day. You have left one far less deserving of Your mercy."

I looked up and the Chinese was standing with his head bowed and his hands folded. He was praying along with me.

I tried to get the chaplain's shoes and jacket but the Chinese would not permit me to do it. I crawled to the porch, dragging my sleeping bag. The Chinese helped me up the steps to the door, and inquired if I was okay. I said, "Yes, thank you very much."

The Chinese laid his hand on my shoulder and said "I am a Christian," and he walked away.

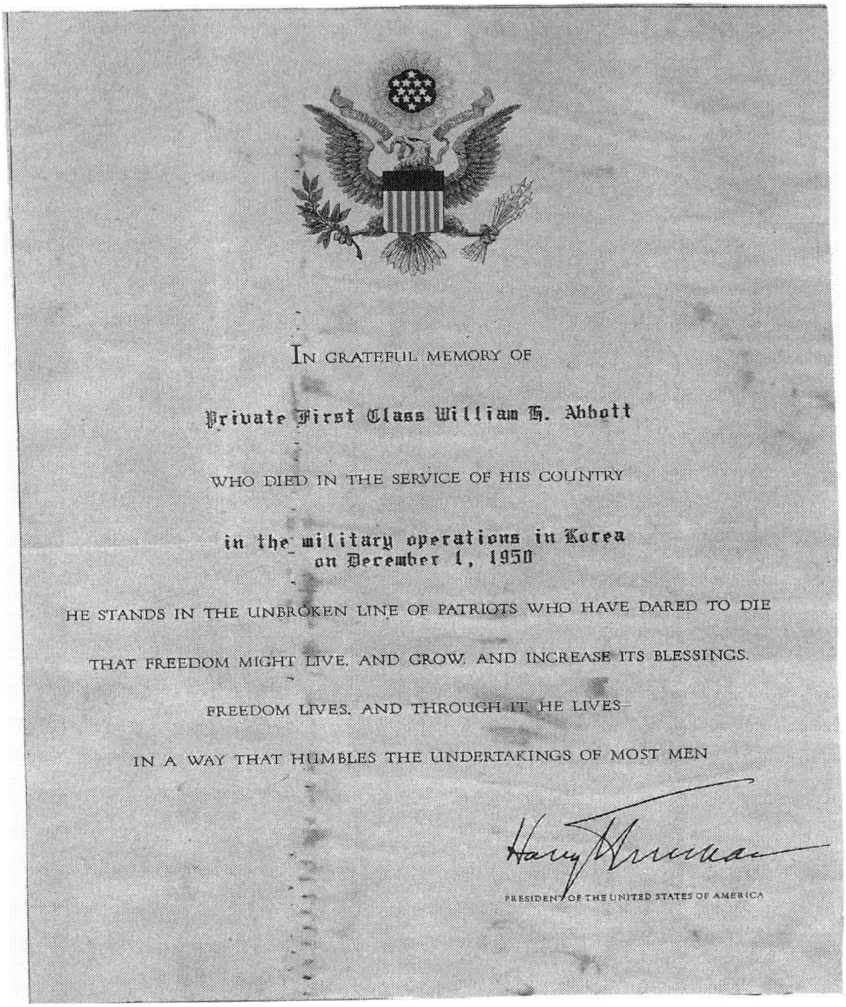

IN GRATEFUL MEMORY OF

Private First Class William H. Abbott

WHO DIED IN THE SERVICE OF HIS COUNTRY

in the military operations in Korea
on December 1, 1950

HE STANDS IN THE UNBROKEN LINE OF PATRIOTS WHO HAVE DARED TO DIE

THAT FREEDOM MIGHT LIVE, AND GROW, AND INCREASE ITS BLESSINGS.

FREEDOM LIVES, AND THROUGH IT, HE LIVES—

IN A WAY THAT HUMBLES THE UNDERTAKINGS OF MOST MEN

Harry Truman

PRESIDENT OF THE UNITED STATES OF AMERICA

William Abbott's Official Death Notice

Chapter 7
Death Valley

I crawled into the room and everyone moved over a little bit. I got in the corner, curled up in my sleeping bag, and went to sleep without saying a word. I did not wake up until the following morning.

The Chinese guard was pulling on my sleeping bag. He handed me four boiled potatoes, all of which were rotten on the outside but good on the inside. I was eating the potatoes and noticed all eyes on me. Only two or three others were eating.

The man next to me asked, “How can you eat those rotten potatoes?”

I said, “It’s simply all I’ve got. I have nothing else to eat. It is not by choice, my friend.”

Another said, “I’d sooner starve than eat that stuff.”

I said, “Suit yourself.”

The others had stored their bran or millet balls up on a ledge and would not eat them. I asked why. They did not know what the balls were. I told them it was the same stuff we fed the cows back on the farm to increase milk production. “If it’s good for livestock, it will help me. If you're not going to eat the bran, don't throw it away. Put it up on the ledge and let me get the nourishment from it." I would lick the ice off and then suck the moisture from the ball, and it eliminated my hunger.

When I needed to go to the latrine, I was not sure I could get off the porch, which was about three feet high. If I did get off, I would have a difficult problem getting back up. I knew I could not count on the others in the room; they were too engrossed in their own self-pity to help anyone. I crawled out on the porch. There were no guards around, only one G.I. standing in the courtyard. I knew the man's face and asked him where he was staying. He said he was staying in the kitchen.

"By yourself?" I asked.

"Yes."

I noticed he was wearing light clothing. He was wearing nothing but a field jacket, hardly something for the cold weather.

"Do you have a parka?" I asked.

"No," he replied.

"Where did you stay last night?"

"I don't know. It was over there in another building. The Chinese brought me here and left me. The kitchen is nice and warm."

"Now listen to me, young man. You are not dressed to withstand the cold weather. It will be ten to fifteen degrees below zero tonight in the kitchen. When that fire goes out, it will cool real fast. You go into one of those rooms where there are other people with body heat, or you will freeze to death."

"Okay," he said.

I crawled over to the corner of the porch, pulled myself up by the corner post, and began to unload my bladder. When a Chinese guard yelled, "No!" and thrust his bayonet into my chest, I fell backwards. As I fell, I continued urinating. He was standing on the ground about three feet in front of me as I fell backwards and the stream hit him directly in the face. He threw his rifle down and began wiping his face with his hands and the sleeve of his jacket.

I crawled back into the room, thinking he had stuck the bayonet deeper into my chest than he had; he had actually just pricked the skin enough to cause heavy bleeding. I did not think I had any blood left in me. It bled heavily but with a little pressure it stopped.

I had to figure a way to get off and on that porch. I might not be so lucky the next time.

The next morning, there was a knock at the door. The man I had talked to about not staying in the kitchen was at the door, calling my name. I opened the paper door. He said, "Abbott, don't be mad with me. I'm froze to death and I shit my britches."

"You dumb shit, I'm not mad with you. Come on in here." I pulled him inside the room and held him in my arms.

"Sure you're not mad with me?"

"No, Son, I'm not mad."

"I'm sorry I disobeyed you, Abbott."

I assured him it was okay, I was not mad at him.

He smiled at me, and then he died.

I did not know this man's name, but he obviously knew me, and it meant something to him that I was not mad at him.

I prayed, "Why me, Lord? What have I done to cause these brave men to have such trust and confidence in me? Whatever it is, Lord, I'm grateful for the ability to help a little. At times like these, Lord, it means a lot."

All eyes were upon me as I struggled to get his shoes off with my left hand. No one offered to help me, but I succeeded. I also got his field jacket. Now I could walk to the outhouse without exposing my feet to the ice and snow.

As I settled down in my corner, I thought, "What a sorry bunch of men this is. Where did such men come from?" These were men who seemed not to care whether they lived or died. They had convinced themselves that death was inevitable, so why prolong it? The sooner death occurred, in their minds, the better.

I looked across the room and there were four men hanging onto each other and an old sergeant, like little chickens and a mama hen with her biddies tucked under her wing to protect them. One young man looked to be about sixteen years old. He had a flesh wound in his leg that was not life-threatening. He was picking at the wound.

I said, "Leave that wound alone, young man. Let it heal or you will get infected and lose your leg and die from that little scratch on your leg."

He looked at the sergeant and the sergeant advised him to leave the wound alone.

Their entire conversation was about food. It was Mama's cooking.

I thought it was interesting. The sergeant was married and had grandchildren, but he made no mention of his wife or other members of his family. He only mentioned his mother. She was the topic of all conversation. But the sergeant wasn't alone, and he wasn't the only one talking about his mother. It was common among the others to talk about the various dishes their mothers would make.

I thought what an awesome responsibility motherhood is. Here these men were, young and old, talking only about what their mothers had done for them.

About two weeks later I had gotten a little bit stronger and was able to go to the outhouse. I opened the door and looked at the sergeant as I crawled out. The he was, with those four young men holding onto him, cuddled up with their arms around him and around each other.

I said, "Sergeant, you've been talking for days about how your dad would beat the hell out of you if he were here because you let yourself get in this situation. Now, Sergeant, if I continue to get stronger and get some usage of my right arm, I'm going to get a stick of firewood out there and when I'm through with you, you'll wish your dad was here because he ain't never put a whuppin' on you like I'm gonna put on you! Look what you're doing to these poor bastards, Sergeant. Can't you see you're killing these young men?"

"How am I killing these boys? I love these boys! I love them like they were my own. I would never hurt them."

"Sergeant, tell me how long it's been since you've eaten or encouraged these men to eat."

"They don't like rotten potatoes or cow feed."

"Dammit, Sergeant, there are no damned apple pies here! Your mother won't be coming through that door to cook your supper. And your dad won't be here to tell you what to do. Now if you want to sit there and die, that's fine. But you will burn in hell for taking these young soldiers with you."

Two days later they were all dead, wrapped in each other's arms.

One of the young men had actually either eaten or pulled the skin off his hands. The skin was hanging down about three inches on his hand, hanging down from the fingertips, showing the bones.

Others died, leaving only a South Korean and me in the room on three different occasions. I asked him, "Who are you, Syngman Rhee?" He said, "No."

I said, "Well, you look like him."

The Chinese brought a large bowl of fried potatoes and set it on the porch, apparently not knowing there were only two men remaining alive in the room. I pigged out on the potatoes and handed the bowl to the Korean. He pointed to where the four men had died the day before and said, "You next!"

I laughed and said, "Let me tell you something, Mr. Syngman Rhee. I will live long enough to drag your dead ass out that door into the snow, and then when spring comes and thaws you out, I will sit on the porch and watch the vultures have lunch! Now we both have something to look forward to. Tell me, Syng, are you mean and cantankerous enough to deprive me of that pleasure?"

We both laughed. He was tough and had a sense of humor. And I knew he was not going to die, even though he had one foot frozen that was rotting off.

One morning before daybreak something hit me in the back. I said, "What was that?"

"My foot," he said.

"What are you talking about?"

He said, "For real, for real!"

I slid the door open to let in some light. Sure enough, it was Syng's old rotten foot. I kicked it out on the porch and crawled over to where he was. He had used a shirt, with one sleeve wrapped around his leg and the other over his stubbed ankle. He had cut off his foot with a single-edge razor blade.

The Chinese guard saw the door open and stuck his head in with a flashlight. He saw Syng's foot and said, "No, no, no, no, no!" and ran out.

Within a few minutes an English-speaking Chinese was there and they carried Syng out.

Syng said it did not hurt when he cut his foot off. He had controlled the blood with a tourniquet made from his shirtsleeve and used a piece of bamboo as a stick.

That was the last time I saw him until just before the war was over when I was part of an "Olympic team," riding on a truck that was passing through the South Korean POW camp. He saw me and was jumping up and down on one foot, waving his crutch. It was so good to see him. I knew he was okay, and I knew he would be okay. He had too much fire in his belly to die.

The room would fill up again with POWs. Rumors began to fly that trucks were coming to take the wounded to the hospital where our wounds would be treated. Two days later several Chinese walked into the room with an American officer. The Chinese spoke perfect English.

"The major has good news for you."

Major Alley came into our room, accompanied by several Chinese interpreters, and introduced himself to us as Major Ron Alley of the 57th Field Artillery. He said the Chinese were making trucks available to us and we would be transported to a hospital and treated.

After assuring us we were going to a hospital to have our wounds taken care of and to get better food, he proceeded to ask questions: name, rank, serial number, all the usual things. Then he began to ask each man for personal information—where he was from, and his family's address. It was inappropriate for an American field officer to solicit non-military information. He asked about our family lives, what our parents did for a living, what kind of work we did before joining the Army, and what, if any, social or fraternal organizations we and our families belonged to.

The major was not addressing me, but I said, "Major, you're going beyond the pale here."

He turned to me and said, "Who in the hell are you?"

I said, "Nobody."

He said, "You're just another smartass."

"That might be, Major, but this smartass will see you court-martialed."

The questioning was over.

A few days later the Chinese notified us that all the men who could walk would be taken to the hospital for treatment. We would have to walk to the trucks.

It turned out there were no trucks. We would have to march to the hospital, and it would be a forced march—meaning a double-time, or rapid step march.

The wound on my back had become infected and was draining down my back. I had developed pain I had not noticed before, and I knew the wound needed to be cleaned. If the drainage turned inside it would kill me. My right shoulder felt like it was healing, although the use of it and my right arm was very limited. I could use it to tie my shoes and put on my clothing without too much pain.

All in all I felt good. God had brought me through one more time.

My left leg was showing the results of constant exposure to the elements. I had not been able to lie on my right side because of the wounds in my back from the Chosin. My left foot had turned black, as was the case with my left knee and my left hip. The skin on my knee had begun to come off, like a big blister, but I could see healthy flesh under the skin. I was beginning to heal from the frostbite. I prayed that my hip and foot would react that favorably. It made no difference. I had to get my back treated. It would be on the march that I would find out how badly my left leg was damaged.

Now my problem was two-fold. I also had to get clothing and shoes that would fit. The shoes I had taken from the young man who had died were much too large for me. My feet were sliding around in them. That would not work. I had no shoes, nor cap, nor parka—three items that were absolutely essential before I could make a long march in sub-zero weather.

There was a soldier about my size who had the clothing I needed. He had a small leg wound that he was sure would prevent him from jumping off the porch and standing at attention as required by the Chinese to make the march north. My attempts to convince him that scratching and picking at the wound would eventually cost him his leg if he didn't stop it fell on deaf ears. He, like the others who were scratching and picking at their wounds, would not admit they were doing it. I said by spring it should be healed, and he agreed.

"By spring, I said, "you will not need this heavy clothing that you have." He had a fear of being left alone without body heat from others. He remembered the young man who had died outside and was afraid he might freeze to death.

"Well let me tell you how we can both benefit. I will trade you my clothing—my combat boots for your cap, and if you give me your boots, boot packs and your parka, I will give you my sleeping bag so you will be comfortable until spring. And if you are isolated, the sleeping bag will end any fear you might have of freezing to death. You will have a field jacket with a new liner that will be more than sufficient to keep you warm."

We made the trade. He was happy with the sleeping bag. He no longer had to sleep on the hard floor.

I had a hard time controlling my emotions. Without his shoes and parka, I knew that I would surely die, but it appeared to me that he was committing suicide by staying there. His wounds appeared not to be life-threatening, and I thought he could have made it to the hospital with the other men on the march.

The cook came to build a fire in the kitchen and I asked if I could dry the liners in my boot packs and my jacket. My jacket had been frozen at one time and was still damp. He had no objections.

The fire felt great, as a fireplace does when you're cold. But it has a hypnotic effect and can quickly disarm you when you're cold. It's easy to fall asleep and wake up after the fire is gone and you're frozen.

My shoes were dry, as was my jacket. I thanked God for His provision and guidance, put on my clothing and my new dry shoes, and went back to my room. The floor was hard without my sleeping bag, but that was okay.

That sleeping bag had saved my life once, and perhaps had done it again. And it had given comfort to the man next door. "Thank you, Sergeant Cutting, for that sleeping bag and for saving my life at the

Chosin," I thought. Then my thoughts turned to the next morning. How was I going to convince the Chinese to let me make the march north? They had let me know in no uncertain terms that they thought I was too sick to make it.

I knew it would be impossible for me to jump off the porch, stop, turn around and stand at attention, do an about-face, put my hands over my head, and say "okay" without falling on my face. I decided to jump off the porch and keep going until I got to the fence, about twelve feet away, and let the fence catch me.

The next morning I got up and got in line with the others who were jumping off the porch. Each man had to qualify to go on the march. When it came my turn, the Chinese said, "No, you cannot go."

"Let me try, dammit! I need to go and get my back taken care of. I can walk okay."

"Okay, you try. You will fall."

I did not move away from the fence until the Chinese walked away. I was a little unstable. The jump off the porch was too much for my left leg, but I reasoned it would be better by morning.

I would not have until morning to rest my leg. That afternoon the Chinese showed up to assemble the men for the march. We went from hut to hut, picking up POWs and lining them up to make the march.

When we were all assembled, it was after dark and the Chinese announced that we would have a forced march of forty miles.

I saw two men I was glad to see: Jerry Francois and Sergeant Cope. I knew I could count on Francois. I had a little less confidence in Cope, but felt secure with him on the march.

We were told that anyone unable to make the march needed to drop out and stay there until spring. A large number dropped out, including Francois and Cope. I was disappointed in them and told them so, but they must have had good reasons for dropping out.

As the column moved out, we walked about three quarters of a mile before we reached the dunes leading up to the reservoir. The dunes seemed to be about three hundred feet straight up and then they dropped off sharply to the reservoir.

As I approached the midway point, my left leg gave way and I fell flat on my face. The rest of the column walked past, some stepping over me without a word from anyone. It was every many for himself; the Chinese had made that clear from the beginning.

After the column had passed, the guard said, “Go back,” and walked on past me.

Something inside kept telling me to get to the top. I crawled until I reached the top of the dune. And there, looking out over the reservoir on my hands and knees, I saw a small white cloud overhead. In that cloud there appeared what looked to me like a woman and a small child. A voice spoke, telling me to stand up and everything would be all right.

The voice spoke again, “Get up!” and I stood up.

The cold wind from the reservoir hitting me in the face seemed to give me a new burst of energy. My leg seemed to be stronger. I could see the column way off in the distance, almost out of sight.

Nighttime in that part of the world at that time of year is almost like daylight so I had no problem seeing the column. I walked and crawled, and got to the reservoir, and then began to walk, following the troops. The further I walked, the stronger my leg seemed to get. I caught the column and moved to the front and led the column all night.

We stopped just before daybreak in a Korean village. We were each fed a large bowl of rice and bedded down until dark. At dark we started to move out again. I could not put weight on my left leg. I started to massage my leg and found a walking stick to help me along with the column. The column once again was leaving me behind. The English-speaking Chinese dropped to the rear of the column to tell me I could not walk, but to stay there and someone would take me back in two days, and he walked away.

I continued to massage my leg, and as the blood started flowing to it I began to walk without the stick once again. Again I caught and led the column. The Chinese reminded me I could not walk forty miles. I said, "Let me tell you something. I can out-walk any son-of-a-bitch in China, and don't you forget it!"

"You might just do that," he said, and he laughed.

I had let my mouth overload my ability. Just before daybreak, I fell out of the column onto the side of the road. My left leg had swollen to a point I thought my pants leg would split open. I was afraid to remove my shoe for fear of not being able to get it back on. The guard kept telling me, "Just a little further just a little further," and he got me a walking stick but it did not help. He walked away saying, "No sleep, no sleep, no sleep."

I could see a small light about five hundred yards up ahead. I pulled the sleeve of my parka down over my hands to keep them out of the snow, and began to crawl, dragging my left leg. My right shoulder and ribcage began to hurt. The pain got heavier with every breath, it seemed. I finally just rolled over in the snow and said, "God, if you're going to take me, you will probably not get a better chance than this."

That same little voice came back again, "You can make it. You must keep moving. Don't go to sleep. You can make it."

I crawled back onto the road again. Somehow the small light seemed much closer this time. As I looked down into the snow, I could see a trail of blood. Many of the men had been wounded in the feet and legs and walking had opened their wounds, and they were bleeding in the snow.

I continued to crawl. By now I was off the road, heading out across the field. The snow was deeper now, making it necessary to rest more often and remove the snow from my parka. I would also warm my hands by putting them close to my body.

When I got to the hut I opened the door and fell inside, lying in the doorway. An old Korean came in with a bowl of rice and a warm cup of water. He lifted my head, sat down on the floor beside me and fed me. I went off to sleep.

When I woke up it was almost dark. I could hear the men moving around outside. The column was forming. When I sat up they were gone. The Chinese guard came back and told me to come when I could.

My first thought, of course, was that I was too far away from the others. No one would know if or when they would shoot you. But the Chinese was very nice to me. He said something to the Korean man and then he left.

The Korean appeared again in just a few minutes. He watched me massage my leg. The swelling had gone down now and I could pull my pants leg up over my knee. The knee no longer felt cold.

I knew that I was getting circulation in my left leg but my left foot became numb. This was always a thought in my mind: don't remove your shoes!

The Korean brought me a bowl of rice and a large cup of hot water. I could not hold the spoon in my right hand. When I raised my right arm the spoon would just slip out of my fingers. But I found I could hold the bowl in my right hand and eat with my left. It worked out fine.

The Korean was pleased. He gave me a good, sturdy walking stick. I thanked him and headed out. He walked with me for a little bit and then turned back.

"I've got to be," I thought, "the luckiest man on the face of this earth. Thank God for His miracle." I made no attempt to keep up with the column but tried to keep it in sight to keep from getting lost. I rested when I thought it was necessary and it worked out well for me.

I followed the column and we finally arrived at our destination. The Chinese had established a receiving station and, true to the Communist cause, it was a staging area, well equipped for propaganda purposes.

Chapter 8
Staging Area South to 38th Parallel

The receiving station was well staffed with those whose sole purpose was to indoctrinate and brainwash the POWs, bandage wounds, and amputate frostbitten legs, feet, arms, or hands—whatever had reached the point of no return.

There were not many whose frostbitten extremities were severe enough to be amputated. As for me, the toes on my left foot and my knee and hip had turned black and the skin was coming off, but there was good, healthy skin replacing the infected skin. I knew my frostbite would heal by itself.

My head wound was healed; my chest wound had a good color and was no longer draining, but the bullet wound in my back was infected and drained down my back like sap from a sugar tree in the springtime.

I was plucking the bamboo out of my skin continually. The bullet wound in my back had penetrated a rib and small pieces of bone were working their way out, not allowing the wound to heal. The Chinese removed some of the bone and the wound began to feel a lot better. I thought perhaps it was healing.

The major thrusts of the Chinese efforts were propaganda and getting all the POWs to sign peace appeals, including and especially the Stockholm Peace Appeal.

After we signed the peace appeals, the Chinese assembled all of us outside and notified us that we would be moving south, and that the more progressive men would be released.

I could not believe the effectiveness of the Chinese propaganda machine and the gullibility of our troops to accept the Communist teachings. The Chinese had found several converts. Eight of them were considered so dedicated they would be taken south and released with the expectation that they would join the Communist Party USA after returning home. In this way they would be more valuable to the Communist Party than as prisoners of war. It would also show that the Chinese were very compassionate toward their captives.

Among the eight POWs to be released was a man named Larson who lived in West Virginia. I wrote a note on a piece of toilet tissue notifying my mother of my health and simply saying I was okay, and gave it to Larson to take to my mother.

I remember Staff Sergeant Olsen from Headquarters Company 3rd Battalion, who was expounding on how great Communism was and how it was the salvation for the world and how the capitalists created all the world's problems. That was the first time I ever heard the term "unjust war."

I said, "Sergeant Olsen, what in the hell is wrong with you, preaching this Communist bullshit?"

He said, "I'm going home, Abbott. And it's damned bastards like you who will stay here and rot."

I busted out laughing and said, "Olsen, I always thought of you as a piece of trash. Now I know you're trash. Not only that, you're stupid trash. Do you think these Chinese can't see through trash like you? They will keep your ass here, work the hell out of you until the war's over, and then send you back home to be court-martialed." And I walked away.

That was the last time I saw Sergeant Olsen. I do know he was not among the POWs released in the spring of 1951.

I was so proud of the men on the march from Death Valley. It was every man for himself, and most carried themselves without complaint. But at the same time I was disappointed at the lack of camaraderie among the POWs. It was every man for himself without regard to his fellow man. One POW could not be expected to help if one of his countrymen dropped out.

Some men were not mentally equipped for such harsh treatment and would ask the Chinese shoot them. I recall one sergeant crying and begging the Chinese to shoot him. That was the second time I had witnessed such an outburst.

I picked up a piece of wood we were chopping there for the woodpile and approached the man and said, "These Chinese will not kill you, but I will. If you continue to disgrace these brave men and your country in front of the Chinese, I'll knock your brains out with this stick of wood."

He got his act together and later thanked me. True to his desire, though, he later died on the march south. It was a sad situation. The man apparently suffocated in his own blood from internal wounds that did not show on the outside. He was obviously in tremendous pain to prefer that his life be taken rather than to suffer.

I looked at the man and bowed my head, and said, "I'm so sorry. So very sorry. May you find peace in God's arms."

The Chinese just left him there beside the tracks. No one knew if they buried him or not. I don't imagine they did.

We marched from dusk to dawn and rested during the day. That night we rode trucks to the railhead, where we boarded a train heading south to the 38th parallel.

I marveled how the Koreans were able to run trains on rubber bands where the tracks had been bombed out. Sometimes the tracks would dip a foot under the weight of the train as we went over the rail, but it stayed together. Luck of the Chinese, I guess.

We arrived at another staging area near the 38th parallel behind enemy lines. On the second day they announced that some would go further south to be released and everyone else would go north. They released eight POWs. We were all excited, with our fingers crossed, thinking that not even the Chinese would move that many men south just to release eight people. We thought surely they would release more.

But the answer soon came. The Chinese assembled us all and said “There will be some going south, and some will be going north.”

“Well,” I told the man next to me, “it looks to me like tomorrow I’ll be going north.” And sure enough, in the morning we headed north.

Chapter 9
38th Parallel Back to Camp 1

The next morning we began our long march north, stopping at POW camps along the Yalu River. This was primarily a march; any other forms of transportation that were used, were used very sparingly.

We traveled at night and stopped in villages during the day. While in the villages the Chinese would line us up and some of the Korean children would spit on us.

Whenever I could I followed the village chickens to their henhouse so I could steal a treat of a raw egg.

We stopped at Camp 5, where the chaplain, Emil Kapaun of the 1st Cavalry, had died the night before our arrival.

The Chinese wanted the Koreans to take responsibility for us but the Koreans knew they could not feed us, and they would shoot us rather than see us starve, so the Chinese took us with them when they left the next morning.

Our march ended when we reached the village where we established Camp 1. We began by gathering and carrying wood four miles back to camp. I was carrying the end of a log on my right shoulder when my entire rib cage felt like it had collapsed. I knew I would not be able to continue on wood detail the next day. Fortunately for me the Chinese announced that night they needed cooks and asked the POWs to select a cook from their troop. After explaining my physical status to the men, they chose me as their cook, which turned out to be a blessing for everyone.

The Chinese brought in a truckload of cracked corn, which we feed the mules back home. The Chinese wanted the corn to be prepared the way rice was prepared, which was not cooked long enough for our digestive tracts to tolerate, and it caused severe bleeding. After much prodding, I convinced the Chinese to allow me to soak the corn overnight before cooking it. I got up a little bit early each morning to be able to cook the corn for a longer period of time.

When the ground began to thaw, wild greens showed their heads. I asked the men going on wood detail to pick them and bring them back to me. I prepared them for the soldiers who were ill with the watercress I found growing in the creek.

Dysentery was killing a lot of men and night blindness was prevalent. Every morning a detail of men carried bodies up the hill for burial, many of whom would be on those litters themselves the following morning.

When the sun went down, the men followed a string tied from their room to the outhouse.

In late spring of 1951 the Chinese opened a potato hole filled with turnip roots and potatoes. We began receiving ten gallons of soybean milk twice a week.

When the water became warm enough, the men were finally allowed to go to the creek to bathe. Everyone boiled their clothing to rid them of lice, which were as big as the end of your little finger. They would embed themselves in the seams of our clothing.

All kitchen facilities were consolidated into one large kitchen area. Officers were moved to a separate company. All non-commissioned officers were also moved to another company. British were separated from American troops, and blacks were separated from whites, but there were two or three black Americans who refused to be placed in the black-only company. They were allowed to stay with the white troops.

Once the segregation was completed, the brainwashing began in earnest. Each group had their own study group and each was given a different course of study.

This was also the time that they were closing down Death Valley and other temporary camps. The POWs remaining in those camps were shipped out to more permanent facilities along the Yalu River.

I looked out the door one day and a truck stopped and dropped off a POW. I thought he must look a great deal like Rip Van Winkle. He hadn't shaved in six or seven months and hadn't bathed properly in that length of time, his hair was way down his back, and he weighed about eighty pounds or so. I looked at him, and I looked and looked. I thought I recognized him, but I wasn't sure until he spoke. Then I recognized him as Jerry François, from Headquarters Company, 3rd Battalion. I walked out and said, "François, is that you?"

He said, "Yes."

I told him, "Come on into the kitchen, I'm going to fatten you up."

I took him in and gave him boiled corn until the other POWs complained about him being there. Of course, that was typical of POWs. They were always worried that someone else was going to get a little bit more than they had.

The Chinese finally came in and told me he had to go, and more importantly, that he had to get cleaned up.

I got François some soap and a towel, and told him he had to go down to the creek and bathe. He said, "No, it's too cold."

When he was complaining about the cold water, I knew it was okay. If he didn't complain, I knew he wouldn't do it. Since he was complaining, I knew he was receptive to doing it. The next thing I knew, he was down at the river bathing. We got him cleaned up, and I didn't see him much after that, until later on when we were separated into groups to move out of that camp.

We formed an inter-company sports competition with basketball and softball teams. I managed the softball team, the Wildcats.

The Chinese made every effort to exploit black-on-white politics in the POW camps. Miles, a black soldier working in the kitchen, informed me one day that he and Hite, another black soldier, were to organize two basketball teams, one white and one black. I didn't think anything of this until I overheard a conversation that on May Day we would march in the parade and the highlight of the festivities would be the basketball game between the blacks versus the whites.

Miles and Hite met in the middle of a field to discuss choosing team members and the rules by which they would play the game. Against the wishes of the Chinese I walked across the field and said to Hite and Miles, "Cameras and news media from the Communist countries and even representatives from the American Communist party will be here to film this event. Miles, you and Hite are good soldiers and much too smart to be used as a propaganda ploy. I hope you can find a way to stop this black-on-white thing. We are all serving the same flag, we have all taken the same oath, and we must honor and be loyal to each other. Our interracial problems are no one's business but our own. Our country will deal with it someday, but right now it is none of the Chinese's business. Why not choose teams with an equal number of blacks and whites on each team?"

Miles and Hite agreed on the mixed teams. When they presented it to the Chinese they turned it down. There was no basketball game after all, confirming my suspicion that it was a propaganda ploy for the media.

A lesson was learned by all that day. The Chinese never tried to exploit the black-versus-white issue again.

Miles and I had previously discussed the color situation in America. I'd had some childhood experiences with our American interracial problems myself.

I had attended a grade school in Ohio where the girl who sat directly in front of me was black. One Saturday afternoon I went to the movies; the theater was full, and the only empty seats in the back of the theater were reserved. I sat down next to the little girl who sat in front of me in school, and she told me that I could not sit there because this section of the theater was reserved for blacks, but I remained seated. Then an usher came and ushered me upstairs to stand and watch the movie until a seat was available.

Miles said he thought the only way the racial issue in America would be resolved would be through intermarriage, simply breeding out black blood. We laughed together and went back to cooking for the men.

Miles was a very strong and loyal man. He respected himself and others, and he would not tolerate anyone degrading him or any other human being.

One morning a black soldier came into the kitchen with a huge rat he had caught. Miles asked him what he was going to do with the rat. He replied that he was going to eat it. We had adequate food and Miles felt he was embarrassing all the black soldiers by eating a rat when it was unnecessary. Miles grabbed him by the throat and said "it has been a long time since I hit a man with my fist, but if you ever come back here I am going to smash your face." The man threw the rat down, took off, and never came back

Miles and I were both cooks. We had never discussed the meeting in the field until one day in the kitchen from out of the blue sky, Miles said, "Abbott, the meeting in the field has never been discussed."

"No, Miles," I said, "because I would be tried for sabotaging the Chinese's indoctrination."

Miles said, "Yes, that is so."

"Thank you, Miles."

Miles started to leave the kitchen and as he reached the door, he turned around and said, "Abbott, how'd you get so damned smart?"

I said, "God made me."

Neither one of us knew I would later be tried by the Chinese for sabotaging their indoctrination.

May Day approached and the Chinese brought out all new uniforms. We were to march carrying the Communist flag. The camp commander, a Chinese colonel, asked me to be the flag bearer for the parade. I refused. He asked why. I responded that it was a matter of loyalty. "I am under oath to my country and I am only authorized to carry the American flag."

The colonel responded, "No problem, Abbott. There are other soldiers who will gladly carry the flag. They have no loyalty to the capitalists."

"Not just the capitalists, Colonel, but to themselves or anything else of value. Colonel, what good is a man who has no loyalty to his oath? The words he speaks fall on deaf ears. He has no substance among honorable men. He is like a ship without a rudder, floating around and around in the ocean with no sense of direction."

The commander said no more. A Chinese soldier carried the flag that May Day.

The Chinese propaganda machine was in full swing. Newspapers from China and the Communist Party USA arrived each week. *The Daily Worker*, published in New York, provided Communist teaching,
as did the *Shanghai News*. My fellow POWs ribbed me for carrying around copies of these newspapers. I was determined to understand the Communist philosophy in an effort to discredit arguments from these papers.

However, the Chinese would be explicit about what they wanted me to say to the other POWs.

We had an outbreak of dysentery that took the lives of many POWs. The Chinese had a fear of dysentery in astronomical proportions and were constantly monitoring the spread of the disease. They set up an aid station in an effort to control the outbreak.

The Chinese treated dysentery with Epsom salts and noodles. Anyone suspected of having dysentery was required to furnish the aid station with a stool sample. However, certain POWs like myself were not allowed to send stool samples because the Chinese did not want us to live. So when I developed dysentery I put the stool sample in a cup and another man's name on the top of the lid and my name under the bottom of the lid. Another soldier carried the sample to the aid station and told the doctor he brought the wrong sample and that it was in fact mine. Within an hour, four Chinese soldiers were there to escort me to the hospital. This, of course, made the company commander furious. I had fallen out of favor with the Chinese because they believed I was sabotaging their indoctrination.

I recovered quickly and went back to cooking.

Before I had fallen from grace due to the episode on May Day, the Chinese gave me a pass to go to town twice a week to obtain the ten gallons of soybean milk from the grinding mill. In those days the Chinese did not watch me very closely. I was able to visit uptown without being stopped. I resumed carrying milk for the kitchen after I recovered from the dysentery.

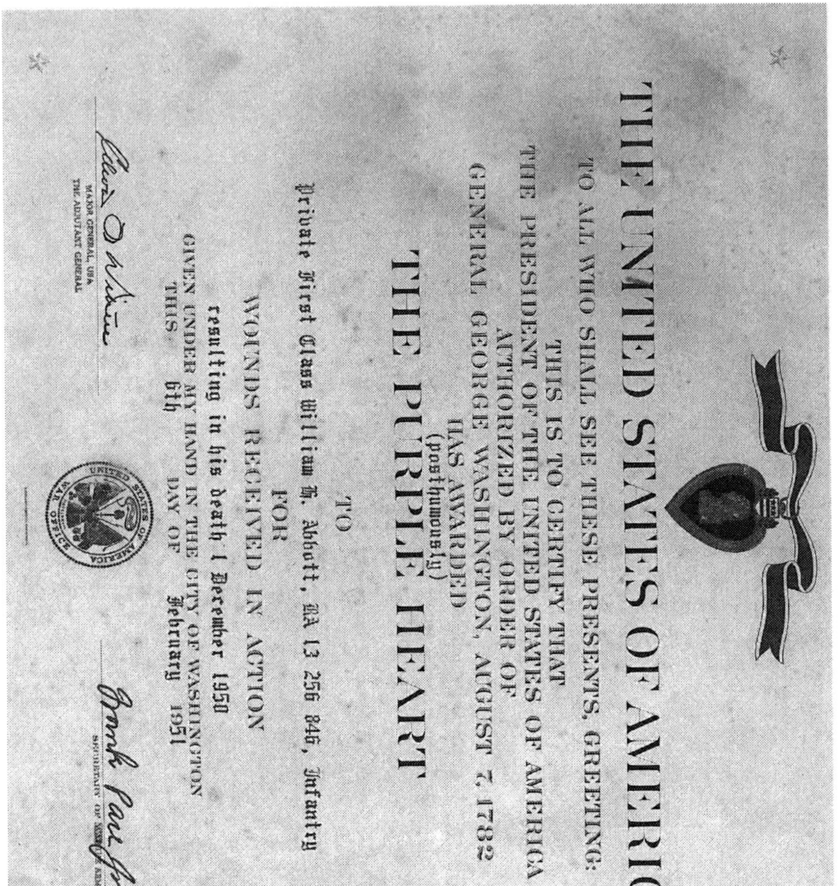
THE UNITED STATES OF AMERICA

TO ALL WHO SHALL SEE THESE PRESENTS, GREETING:

THIS IS TO CERTIFY THAT
THE PRESIDENT OF THE UNITED STATES OF AMERICA
AUTHORIZED BY ORDER OF
GENERAL GEORGE WASHINGTON, AUGUST 7, 1782
HAS AWARDED
(posthumously)

THE PURPLE HEART

TO

Private First Class William H. Abbott, RA 13 256 846, Infantry

FOR

WOUNDS RECEIVED IN ACTION

resulting in his death 1 December 1950

GIVEN UNDER MY HAND IN THE CITY OF WASHINGTON
THIS 6th DAY OF February 1951

MAJOR GENERAL, USA
THE ADJUTANT GENERAL

SECRETARY OF [illegible] ARMY

Chapter 10
Camp One

I tried to arrive at the mill a little early each morning to talk to a young woman named Kim, who directed the distribution of the milk. Our conversations were short. She wanted to know about life in America, New York in particular, and Coney Island. Al Capone, the gangster, fascinated her. She was reluctant to talk about Korea under Kim Il-Sung's Communist government. Things were hard for her and her family after the Japanese left in 1945. The Russians were mean to the Korean people.

Times had been good under the Japanese. The Koreans never experienced a shortage of food, shelter or clothing, and the Japanese increased rice and cotton production. Kim would add the Communist line, "They were imperialists and could not be trusted. Their first loyalty was to Japan, not Korea."

"But you just said, Kim," I remarked, "under Japanese rule and occupation, things were good."

"Oh yes, good, and more free!"

Each day Kim became a little more open with her conversation. She talked about her mother and father riding the train to visit relatives in Pusan. It was clear she thought growing up under Japanese occupation wasn't terrible. As a child she went to school, worked hard in the fields, and had plenty of playtime. Under Russian rule and Kim Il-Sung, there was never time to play. "The food went to the Army and the people got what was left over," she said. With Communism, all is committed to the common good.

One morning the wind was blowing cold and very hard. The mill was running out of soybeans. The Korean in charge of the mill came over to Kim and told her about the soybean shortage. He told her that it would be perhaps two hours before they could get more soybeans. The Chinese had requested more milk for new troops arriving from the north.

Kim asked the Korean if she could use his house to keep out of the cold until they returned. He said yes. By this time the ox had been unhooked from the mill and hooked to the cart to carry the soybeans. Everyone left the mill and she invited me to join her in the house.

I followed her to the house without speaking a word. The room was very warm. She walked across the room, turned around, and said, "One hour," and ran into my arms, removing her clothing as she approached.

"Kim," I said, "When two lonely people need each other, and can meet like you and I are doing now, one hour is a lifetime. Let us treasure every minute."

Time went quickly. I could hear the old man hooking the oxen back up to the mill.

We walked down to the mill. I got my milk, and headed back to the company.

From that day forward, we arrived before the old man left the house for the mill. She would bring a bowl of rice for him. We would stay warm until the mill was in full production, which was usually between forty-five minutes to an hour and a half.

One morning Kim told me she was to accompany the old man to the river to purchase soybeans. The next morning she talked excitedly about the trip.

I said, "I bet it took you a long time to get there and back with all the roadblocks."

"No roadblocks," she said, and that was the beginning of my escape attempt.

If there were no roadblocks, the Chinese felt secure, especially if it was not necessary to guard the road. If I could steal a truck in the early morning after the mill had been closed, no one would miss the truck before daybreak. And with luck, the Koreans would think the Chinese had taken the truck. I could make it to the river, steal a fishing boat, and go out with the tide. I began to store food.

I knew that everyone that tried to escape had been caught while stealing food.

The bounty to civilians was fifty pounds of rice for the capture of each POW. For people suffering from starvation, that fifty pounds of rice could make the difference between living and dying.

I had to avoid the trap of being caught. Hard winter began setting in and now everything would have to be put on hold until spring.

The British and French had a way of creating entertainment for themselves and others. It was nearing Christmas, and the Chinese had given them permission to put on a Christmas play. The troops were permitted to go and rehearse in the town theater.

One morning as I was returning from the soybean mill, the theater group was heading uptown for a rehearsal. I asked if I could join the group.

"Oh, yeah!" they replied, and that afternoon I was at the theater, singing and dancing and flat "putting on the dog."

The first night of the production was for the village population. The second night was for the POWs from the officers' company, and the theater filled with high-ranking North Korean officers as well as the American officers. When the play was over, a Korean officer approached me and asked if I spoke Japanese.

"Some," I said.

The Korean introduced himself as a colonel stationed in Pyongyang, and said he was there to visit his mother. He asked how things were going.

"Okay," I said.

He asked about food, and how we were being treated. I told him the food had improved.

"Have you been told of the new peace offerings that have been made by the People's Republic to bring the war to an end and send us home?" he asked. "If the imperialists will accept our most generous offer, you could go home in three months."

I thanked him for the good news. "I wish you and yours a prosperous and healthy new year, Colonel. I'll pray that common sense prevails at the peace talks." I bowed, and shook his hand. "Thank you again, Colonel," I said.

One of the Americans grabbed me and wanted to know what the Colonel had told me. "Nothing of substance," I said, and I left the theater. As I passed the colonel, he asked, "Why did you not tell him what I said?"

"Your message was a good Christmas present, Colonel. I didn't want to spoil it with details." He nodded and walked away.

Our final performance was on Christmas Eve. The house was packed with senior Korean and Chinese officers and people from the village. There were no young men or women in town except medical staff, schoolteachers, and those essential to the government operations, such as the girl running the projector and managing the theater. She had filmed the entire play for the Chinese government.

After the play was over, the Chinese and Korean officers stood around talking. The elderly sat patiently waiting for their turn to leave. After the officers had left, the theater manager signaled it was time to move out, and the old people left. Our theater group was the last to leave the theater. We followed the old townspeople out the door.

A young Korean boy of about sixteen years of age came up beside me and said, "She likes you. You stay," and he pointed to the woman running the projector. She was actually the theater manager. As I walked out, she was standing at the door, ready to lock up. I stepped away from the others and walked behind the door. She locked up, and I followed her home.

There were Russians in town, and you could hear them from about three blocks away. The Russians were boisterous and loud when they were drinking vodka. We walked several extra blocks to avoid contact with them.

She lived with her grandparents, who were quite old. She did not know the whereabouts of her husband and her sixteen-year-old son.

I removed my shoes in respect for the Korean custom; this gesture helped Papa-san and Mama-san relax a little. The house was large, with a kitchen, a large entertainment area, and twelve small rooms. I said "Hello" to her grandparents, and shook her grandfather's hand.

She told her grandfather I was an American. He said "Number One." He had thought I was a Russian. He reminded me that I could get into a lot of trouble, as well as cause them a lot of trouble. He wanted to protect his wife and his granddaughter. His granddaughter assured us all that it would be okay. She chose one of the rooms in the rear of the house for us to spend time together, assuring us some privacy.

After about thirty minutes we heard loud voices outside. She jumped up, "Your shoes!" she said, and ran up front to hide my shoes. "You go," she said. She slipped the back door open, then came back and said, "You stay."

She turned, ran to the front door, came back with my shoes, and said, "Come quickly. We go."

When she said, "We go," I breathed a little stronger. She led me back to the theater, from where I knew the road back to Camp. I traded a GI watch for a bottle of sake and placed it in the sleeve of my overcoat.

When I arrived at the camp, the men were sitting on the porch like a bunch of crows sitting on a fence, not speaking a word.

“What on earth is wrong here? Why all of these long faces when you have so much to be grateful for?”

“Abbott, what have we got to be grateful for?”

“Well, for starters, you have a full belly. For another, the one whose birthday we're celebrating today has chosen to give you life, while others are lying buried up on that hill.” I ended up preaching a full sermon that night on the birth of Christ.

One man said, “Abbott, you're as drunk as a skunk, and preaching.”

The man next to him said, “We’re very lucky to have a man like Abbott, who can preach a sermon drunk or sober.”

We sang Christmas carols until two o'clock in the morning. We were joined by a large number of English-speaking Chinese. No one asked where the sake came from.

The next day I went back to cooking for the troops, and continued storing food for the escape I had planned in the spring.

One of the POWs I had helped during the march across Korea, Edwards, turned me in to the Chinese for storing food. Each meal I would take and hold half a handful of corn or rice. When the Chinese found the food my answer was, "The sick needed extra rations for recovery."

"If the sick need extra food the doctor would ask for it," they replied. I would not be able to store food for those wanting to escape. It would be several months before I would be in good enough relations with the Chinese to store food again and had sufficient supply to escape.

My plan was not foolproof but I thought it stood a good chance of success.

The following month I gathered information on the trucks, the railhead, and the boat dock. If Kim was correct, the Chinese considered the road secure enough to be unguarded. Once at the dock we would hide until high tide, capture a boat, and sail to the open water. Kim had said American boats were operating just off shore.

I needed four men to ride the truck with me, to avoid suspicion if there were fewer men. Where would I get four dedicated men who would keep their mouths shut in a POW setting? Secrecy is next to impossible to maintain among POWs.

I found four men that I had confidence in and we began gathering food for our escape.

By the time we gathered enough food the rains had come, making it difficult to walk in the muddy soil, but it also meant we would not be detected by the Chinese guards as easily.

They would cross the river while I went to get the truck, using my pass, and I would pick them up at the road leading to the railhead.

Our attempt to cross the river at the creek by holding onto the top of the homemade dam was disastrous. Had we gone across in single file, the plan would have worked well, but all five of us going across together was too much for the little homemade dam to hold. When we reached the center of the creek, the dam broke, sending us down the river.

I grabbed a tree branch. Leo came flying by. “Abbott, help me! I can't swim!” he said.

“Just keep doing what you're doing, and it will be okay.”

He was swept about a hundred feet and then managed to catch onto a tree limb and climb out of the water.

Pat Patterson was the last to swim out. We had lost all of our food. I decided to try walking around the steep hill on the south side of the camp to try to reach the truck. The truck was the part of the plan I had never discussed with anybody. We stopped to rest and I realized that we had less than two hours before daylight. We were out of food, and if we did reach the truck, it would be too late. The mill operators would be cranking up the mill, making it impossible to steal the truck. I decided our only option was to return to camp and try again another day.

We worked our way back to the camp. Pat Patterson was complaining that he and I should continue on. He didn't know what he was talking about.

Later I lay there in my warm bed, thinking about next time. We would try to do better.

That night, when two guards poked their heads through the door and shined a light in my eyes, I knew that someone had already told the Chinese about our escape attempt. It was no secret any longer.

I was soon to find out that Roy Hardage, another POW, had notified the Chinese of our escape within minutes of our departure. Our plan was doomed from the start. Hardage was a good friend of Leo, and Leo had confided in him. The Chinese arrested Leo immediately.

Until now the Chinese had been unable to arrest Leo. He would not confess to any crime, but now they had him and he was in a Korean jail.

The Chinese wanted a ringleader, someone to blame, and they chose me.

At daybreak I was arrested and they had me stand at attention for ten hours.

At night the temperature would drop to freezing, and rise to forty or fifty degrees during the day. This was one of those cold, windy days. The temperature hung below zero, and my shoes were frozen to the ground.

A Chinese soldier, Lieutenant Wu, read the names of all the men who were with me who had tried to escape. "Do you want these men to suffer for your stupidity, Abbott?"

"They are grown men, Mr. Wu."

"Abbott, you are the leader. The commander only wants to punish the ringleader. You confess, the others will go free."

I said, "Okay, I forced the men to go."

He wanted me to write a confession.

"I am too cold to write. Give me two hours."

"Okay," he said, and the guards took me to the kitchen. I fell asleep there, and was awakened by Wu telling me to write the confession. That would provide the commander a reason to punish me, and an excuse to free the other men. That would make him look good.

"Those men would never even consider leaving the camp without pressure from Abbott," the commander said.

I agreed to serve any additional time in place of the others, whatever time the Chinese deemed necessary to compensate for the other men. I knew full well that the Chinese were going to punish me, and the other men had nothing to do with it. I was glad it worked out that way.

I told the men, "Just keep your mouths shut." They already had me and things were not going to change.

I was turned over to the North Korean government and placed in a Korean jail. The jail had a concrete floor and the bars were made out of four-by-four-inch wood. There were no steel bars and no sleeping provisions. The prisoners sat cross-legged at attention, arms folded, and eyes straight ahead. When nighttime came, you would simply lay your head back into the lap of the man behind you and go to sleep.

Anyone caught not sitting at attention would be punished. The man would be placed front and center, standing on one foot, arms outstretched in front of the door. The door would be opened just enough for the man to put his head through.

The latrine was located at the far end of the jail. Upon latrine call, every prisoner would slap you as they walked by to the latrine. Most would only slap their hands to make it sound like they were slapping you hard. But there were those who were trying to impress the sergeant of the guard, and they would hit you so hard you would go flying across the cell.

The Koreans had another little trick. They would send a young Korean into the room as a fake prisoner. His job was to slip up behind a prisoner and slap him as hard as he could.

There's a strange thing about being slapped. After a while, your face becomes numb. You don't feel a thing. Oh, you see the stars and feel the blow, but no real pain. I recall being hit so hard by a Korean once that the imprint of his hand remained in my face for well over an hour.

The Korean guards would also harass you with slander against President Truman and General MacArthur, your country, your family, trying to get you to react, which would give them an excuse to punish you more.

On one occasion a crusty old Marine, Sergeant Pettit, who was a POW of the Japanese during World War II just across the river in China, was asked if he and MacArthur were the same. Pettit replied, "Yes."

The Korean guards jumped on him and put him in the door. When it came my time to slap him, I refused. "The man is sick," I said, "I'll not slap him."

"He and MacArthur are the same. Warmongers. Are you and MacArthur the same?" the guard asked.

"No. I'm not worthy of shining General MacArthur's shoes"

"You come," they said.

Sergeant Pettit sat down. I took his place at the door. At latrine call the door opened and I put my head out. The prisoners proceeded to slap me. It didn't take much of a blow to knock you down when you are standing on one foot with both arms outstretched.

The food was a small bowl of bran, a large spoon of soybean sauce, which was about forty percent salt, and a small cup of water. I lost a lot of weight.

One day the sergeant of the guard called me to the front of the cell. He wanted to play a game. The Koreans had a game where a line would be drawn on the floor and each would try to pull the other man across the line. The sergeant of the guard knew that I was too weak to pull. His objective was to pull me through the bars and split my head open or smash my face. The bullet wound in my back was infected, and the infection drained every ounce of energy from my body.

I put out my hand to pull. He had called out everyone to watch me. The bars were made of four-by-four rough-cut lumber and the edges were very, very sharp. I could get hurt very badly. I said, "God, I sure need your help this morning."

The sergeant of the guard was showing off, getting attention. Before he could get his feet set to pull, I could see he was off balance. I pulled him with every ounce of strength I could muster. His head hit the bars and blood came flowing down his face. His nose was bleeding. The sergeant ran back to his office and I sat back down.

One man said, "You will be shot."

I was pleasantly surprised that nothing further happened that day. The following morning, the sergeant came back on duty. I could see his eye was swollen, and his nose had a heavy cut on it. He called me out in the hallway, and he jumped down in pulling position. He put his hand out and said, "Pull!"

I reached out my hand to pull, and a big smile came on his face. He knew he could hurt me badly, and would look good in front of his men. Still smiling, he did not pull. He looked me square in the eye, and said, "You are Number One."

I shook his hand, bowed, thanked him in Japanese, and went back into the cell. From that day forward, the sergeant would nod his head whenever he saw me. This was his way of saying "good morning."

I remembered the advice of my old Judo instructor at Camp Crawford. "Never let your opponent see fear in your eyes. It is a sign of weakness, and it gives your opponent confidence. Never give him that edge under any circumstances, Abbott."

I may have unconsciously taken his advice. Whatever had happened, I thank God for the way it ended.

The Chinese began their interrogation. It was cruel. They would sit you in the bright sunlight as you shivered uncontrollably. After thirty days underground the bright sunlight would momentarily send your body into shock.

"If you want me to answer your questions, Comrade, let us go inside out of the sun." He laughed and agreed.

"The sergeant of the guard tells me you're a model prisoner, Abbott, and he says you are a strong and honorable man."

"That's very kind of him, Comrade. I believe him to be a good soldier, and an honorable soldier."

"I'll tell him that. Now then, Abbott, You have committed crimes against the Chinese and Korean people. You must confess to all of your crimes. Do you understand that, Abbott?"

"Yes, Comrade."

My confession was simple. I had cut down one of the people's pine trees. It was a lie, but I had confessed to that before. I had violated camp rules by attempting to escape, and forcing others to go with me. Now it is my duty to go before the entire POW population and tell them what an exercise in futility it was for anyone to try to escape.

Punishment by the Chinese and Korean People's Army had been just. Thirty days' hard labor was a just and fair treatment for bringing hardship upon the other men, and I signed the confession.

As we walked back to the jail, we passed a man's body that had been lying there for about two weeks. I asked, "Comrade, when are you going to give that man a decent burial? His body is beginning to rot."

He laughed. He would be buried in due time. I realized the man's body was left there on purpose, freezing at night, and thawing out in the daytime. The smell from the body was part of our punishment. I thought, "You sick bastards." That old Korean had been given fifteen years' hard labor working in a salt mine. He had held back fifty pounds of rice and made it into sake. Now, too old and too sick to work, he was sent home to die. He found an empty sake bottle in the latrine that had probably been left there by one of the guards, and committed suicide by breaking it and jamming it into the top of his head.

Three days later the Chinese came and took me from the jail. We walked to the parade field where most of the POWs had been assembled. The Chinese read off the list of punishments for the escape, and announced I still had thirty days of hard labor to be completed. I gave a speech telling the men how difficult it was to escape, and how we should be grateful for the Chinese hospitality.

"Yes," I said, "it is every soldier's duty to try to escape if possible, but I caution you, Gentlemen, think long and hard before you attempt such an escape. Thank you, Godspeed, and God bless you."

A guard marched me to camp headquarters, where there was a mountain of firewood waiting to be split. I had thirty days to finish splitting the wood.

I was assigned to a squad of men who had not been allowed to associate with the general POW population. They were a mixed bag of people. Some were English, some Americans, and some Arabs. For some reason I never quite understood, we were not allowed to speak to them. I was not allowed to speak to anyone in the squad. When I was a cook I would cook their food and set it on the porch and leave.

There were all sorts of rumors floating around about that squad. Some would say they are going to be released for "special study groups." Others would say they were going to a labor camp. I was told not to associate with them. I was to sleep there, cut wood from sunup to sundown, and I would continue the hard labor for thirty days or until that pile of wood was gone.

My left hand had begun to get infected; it was badly bruised from swinging the ax for thirty days.

A Chinese colonel sent for me. He asked me to come into his office. He wondered where I had learned to split wood and use an ax like that. I explained to him that my family cooked on a wood stove and we heated our house with wood in the fireplace. We also cut and sold wood to the rich folks downtown. He laughed.

"Abbott," he said, "these old Korean farmers are having a hard time gathering their crops. Would you like to help them?"

"No, Sir. In all due respect, Colonel, I cannot justify sending North Korean soldiers south to kill our troops. If the shoe were on the other foot, Colonel, I would not expect you to do it. Now Colonel, if those were Chinese farmers, I would feel differently about it. Because you, Sir, are feeding me, and I feel obligated to pay for my food and lodgings."

"Abbott, would you explain what you mean by 'shoe on the other foot'?"

"Yes, Colonel, if you were in my position, and I in yours, I would not ask you to do something that would provide troops to shore up the front lines and fight against your people."

"You can return to your company now, Abbott." He told the guards to take me back. I thanked the Colonel, turned and walked back to finish chopping up the small amount of wood left in the woodpile.

The Colonel said, "Abbott," as I reached the door. "I am proud to have met you."

"Thank you, Sir."

"And I am sure that your country is proud of you as well."

"Thank you, Colonel. It is very nice of you, and very kind of you to say that. This old world will always remain a better place as long as there are men in leadership like you, Sir. You have inflated my old Southern pride this day." I bowed, he smiled, and I walked back to work.

I walked across the yard, which at one time had been a large woodpile.

The guard there was anxious to try out his new English language course he was taking. There was someone talking over the loudspeaker in English.

"Do you know him? He name Bates. He got bad heart."

"No, I do not know him."

It took me about an hour to finish cleaning up the woodpile. As we started back to the squad to pick up my food bowl and spoon, where John Kirby, Contie, Martin and others were staying, two Chinese interpreters and a heavy-built American POW were walking down the road.

"He Bates," the guard said.

I arrived back at the company and the company commander told me I was not to associate with the other POWs. I asked the guard what he thought was going to happen to the men in the squad I had just left.

"They go labor camp," and he pointed across the Yalu River to China.

The company commander expected me to bring back a lot of wood when we went out on wood detail. I assured him I would.

Within two weeks, my feet began to swell, my left hand became seriously infected and was turning blue, and severe pain was shooting up my arm. I borrowed a razor blade and tried to cut down to the infection, but it was too deep in my hand. I went down to see the company commander about my hand.

"I have no sympathy for you, Abbott."

"Damn it, Captain, I don't want your sympathy. I want my hand lanced and the infection removed!"

"Put your hand out here, Abbott. I'll lance it for you." He pulled out his pocketknife, and began to pull at my hand. He did not cut into the hand and sent me back to my room.

Soon Mr. Wu showed up. He wanted to talk to me.

"Abbott, make no mistake about it, these people intend to kill you. I think you're too smart to let that happen. You do not have much time until gangrene sets in that hand and then it will be too late."

I said, "What do you suggest I do?"

He fell silent. Then he said, "Let me see your back."

I pulled up my shirt and it must have smelled pretty bad. He grabbed his nose and motioned for me to put my shirt down and said, "You have one choice."

I asked what that might be.

"You have to get progressive."

He then proceeded to tell me about how he came to be in Korea. He had been a major in Chiang Kai-Shek's army prior to the Communist takeover by Mao Tse-Tung.

He was working at the university when he was asked to volunteer to come to Korea as an interpreter. He refused. He was told in no uncertain terms that if he valued his family he would go to Korea. "So here I am," he said.

"And I thought I had troubles."

He pointed out that the company commander was convinced that I was a ringleader and was sabotaging their indoctrination among the POWs. He said the commander thought I was the one who embarrassed him when I encouraged Miles and Hite, two black soldiers, not to arrange inter-camp competition between the blacks and the whites. He was, of course, correct. I had encouraged Miles and Hite to have two mixed teams of equal numbers of black and white on each team.

The camp commander was convinced that I was a strong man with strong convictions, and he would be pleased if I became progressive.

I said, "Okay, what do you want me to do, and when will I get to see the doctor?"

"If you will agree to teach the other POWs about the values of Communism, I'll get you to a doctor."

I said, "Okay. But when will I get to see the doctor, and when will I get the information the commander wants me to convey to the men?" He would tell me what to say. We agreed, and we went on our separate ways.

Within the hour, six Chinese soldiers showed up with a stretcher to take me to the hospital, which was actually nothing more than an aid station.

The doctor was a young Chinese woman who had gotten her medical training in England and spoke with an English accent. She looked at my hand and said, "You should have been here sooner. The infection has spread to your elbow and up the arm. Your arm might need to be amputated if we are unable to stop this infection."

She attempted to numb my hand, but the novocaine came back out through the open cuts caused by the razor blade I had used to relieve the pressure on my hand. She cut between all my fingers and then inserted a wire into my hand until the infection was running out between my fingers. She then directed the nurse to insert a drain into my hand.

The Chinese nurse was a young girl, perhaps twelve or fourteen years old, and as she forced the wire back into my hand, she must have hit an artery, as blood and pus shot out and hit her in the face, which caused her to jump. As she moved, the wire she was holding went through the back of my hand. I thought my hand was on fire. The doctor heard the commotion and told me to hold my hand over my head. Then she covered my hand with black salve and dressed it and told me she wanted to see me the next day.

She then told me to remove my clothing down to my underwear. She wanted to see my back, and told me to lie down and then said something to the Chinese, apparently giving the guards instruction on what to do.

One of the guards sat on the back of my neck, holding my head and my feet while she opened the wound and pulled out two pieces of what appeared to be bone, part of the rib. She then proceeded to open the cyst on my tailbone, which she said was affecting my legs, and again packed it with salve. This salve looked like the same stuff we used for the horses' feet back on the farm.

She told me to sit in hot water and soak my hand if I could.

When I got back to the company, I crawled into the kitchen and told Pat, the cook, to add extra water to his rice water and give it to me. Within three days the swelling was gone from my hand and the color was coming back.

When the doctor removed the bandage from the bullet wound, she said it was healing. After applying a new bandage, she said it would be okay to remove it in three days. I told her God had chosen her to be a miracle worker, and I thanked her.

When I returned to the company, my old trusted friend Wu was waiting for me. He had three issues of *The Daily Worker*, a Communist newspaper published in New York, and a stack of other material. I was about to get a quality education in Communism, directly from the master professor himself.

I was taken off wood detail and allowed to spend full-time studying Marx, Chairman Mao, the Soviet Union, Joseph Stalin, and world Communism. I was a prime pupil.

In the evening, when all of the POWs were sitting around outside, Wu would have me walk with him in the front of all the men to show off his prime bull.

A remark had been made by one of the Chinese commanders that my blood ran red, white, and blue, and I would stand on my convictions.

Converting me was more valuable to him than converting the entire company. I put everything I had into studying Marxism and Communism and the world Communist movement from 1917 to 1950. I became so knowledgeable in the subject I was called Gus Hall, referring to the Secretary General of the Communist Party USA.

I was comfortable discussing the revolution, and the genius of the great Mao Tse-Tung and his defeat of the Japanese; the subsequent defeat of the capitalist American puppet, Chiang Kai-shek; Franklin Delano Roosevelt, the great friend of people all over the world; the great liberators of the world's people, Karl Marx; Friedrich Engels, Vladimir Lenin, and Joseph Stalin; and, of course, the warmongers, Truman and MacArthur.

Not until I began studying Lenin's obsession with starting a revolution in America by exploiting the American Negro did I fully understand why the Chinese would not allow black POWs to associate with the white population.

The American Labor Movement during the 1930s was a special topic of interest to me. I was amazed at how well the Communists had done in American labor. The support of the Communists by those within the Roosevelt administration literally blew my mind.

I enjoyed these discussions with the Chinese interpreters. They would open up every facet of the teaching that had been taught to them before coming to Korea, letting them know precisely what to say to the POWs when in their presence.

It was obvious these young Communists had been exposed to only one side of the argument. These were bright young men who could see through some of the teaching as purely what it was—propaganda, designed for those less educated and informed. There was a training film showing how the Soviet Union had single-handedly defeated Hitler. Some of the tanks used in the film were obviously American tanks. No one talked about that.

My discussions with the Chinese staff would prove to be better than those I conducted with the POW population. POWs had learned their lessons well and were always happy until in conversation with the company commander. They had repeated what I had said about Communism, and about an individual's ability to advance himself according to his ability and achievements as an individual and not rely upon any government to determine what his needs and desires were. I had said that under Communism there would be a jailer-inmate relationship as it relates to the individual's freedom. All forms of collectivism, whether it's Marxism, socialism, or whatever titled, all are incompatible with the free spirit of man. The god of materialism as conceived by Marx is a very strong and demanding god, and that god requires full submission to the state.

I should have been much smarter than that. I knew better and should have stayed clear of that trap.

The Chinese decided they had created a monster. The one person they thought to have been converted as an enthusiastic Communist that would join the Communist party when the war was over had, in fact, used them, and was now in a position to sabotage their indoctrination.

The Chinese knew, and we had mutually agreed, that my position was a soldier under oath. Right or wrong I would not violate that oath, and they would not ask me to violate that oath. In their view, they had kept their side of the bargain but I had not. I was unable to convince them, and I was in very serious trouble.

I had said that man is spiritual in nature and is always searching for God's help to find fulfillment in life. From the beginning of time, man has served his god. History has shown that man desires to assume responsibility for his actions and choose his own future. It is as natural as the air he breathes.

God, in one form or another, has always been man's refuge. Every other nation has relied on its god for guidance—Muslim, Buddhist, Hindu, or Karl Marx, a god of dialectic materialism, the man-made god of the state—all require total submission of the individual's rights, and therefore are at the opposite end of the spectrum from the God of the Bible.

The Bible makes it clear that man is responsible for his own choices. In America all laws flow from the well of Christianity. America's belief in the God of the Bible, its laws, its institutions of government, and its citizens' individual responsibility are based upon the Ten Commandments of God handed down to Moses. The Ten Commandments are ten simple rules, guidelines if you will, that have led America to be one of the most prosperous free nations on the face of this earth—not because of government, but because the Bible teaches the responsibility of one's actions.

I was placed in an old building, called a coal box, and stripped down to my underwear, but the guard allowed me to keep my jacket and shirt. The temperature dropped down to the low twenties at night and would go up in the daytime. The Chinese insisted I sit on the floor with my feet down in the flue holes.

All buildings in that part of the world were built with a chimney and a sub-floor. That is how they heated the house. There was a very primitive fireplace outside that would heat the rocks in the chimney which were covered with mud and that were also part of the finished floor. The system worked well if used as it was intended. When buildings were neglected, the mud floors collapsed, leaving access to the outside. Also, without maintenance, the walls would develop large holes and the mud would fall away from the bamboo supports, leaving it open to the outside.

That was the situation in which I found myself. The sub-floor had crumbled and large holes had opened up in the floor and outside walls. I used the jacket to wrap my feet and when I felt myself falling asleep, I would jump up and run around the room until I had worked up a sweat and was exhausted and then I would sit back down and rest. The guard heard me running around the room and ordered me to sit down or he would shoot me.

I said, "Shoot." He walked out of the room and got the interpreter. The interpreter said I was to sit at attention or the guard would shoot me. I said, "Well, if I'm going to die, you'll have to shoot me. I'll not make it look like I froze to death or, worse, have my feet freeze off and die from that." He ordered the guard to fix his bayonet. He said he would talk with the commander.

By the time the commander arrived, the troops in the other building were awake and several came over to see what was going on and why I was running around the room. It was my intention to let the troops know what the Chinese were doing. I had no doubt that without witnesses they would have let me freeze to death; at the very least my feet and legs would be severely frostbitten, but neither happened.

The order was given to allow me to return to my company. The men used every blanket they had to wrap me up and thaw me out. I was much closer to freezing to death than I had realized.

I thawed out after about two hours. I had stopped shaking and was recovering just as the commander wanted to see me again. This time I was ready for the cold. I had three blankets, two pairs of socks and a heavy coat. He saw the amount of clothing I had on and laughed. He did not send me back to the coal box.

One thing about the Chinese Communists was that once you had signed a confession, everyone who wanted be a commissar looked upon you as fair game for interrogation. If they were able to get you to sign a confession for a more serious crime than your previous confession, they could simultaneously build themselves up while breaking you down.

A man named Sergeant Pettit, who had been a POW with the Japanese in World War II just a few miles from where we were there on the Yalu, remarked that he would rather spend five years under the Japanese than one with the Chinese. According to him, if you disobeyed an order with the Japanese they would knock the hell out of you (his words) and it was forgotten. With the Chinese, it is never forgotten. They keep using the same old excuse to punish you, over and over and over again.

Actually, there may not even have been a violation that they asked us to confess to; a man may confess to a false violation just to keep from being harassed. A case in point: I once confessed to cutting down one of the people's pine trees while on wood detail, which was a lie. The punishment was a reprimand in front of the whole company, which was much better for me than standing at attention in the cold. The confession made everyone happy. The Chinese interpreter could show he was a good Communist by being vigilant and doing his job, and I was able to get in out of the cold. That is the kind of game they play. Spying on each other is a way of life. That is how promotions are gotten within the society as a whole, not just in the army. It is said that one in ten workers is a government agent under the Communist system.

The night without clothing in the coal box had taken its toll on me. It seemed everything was going wrong. Physically, I could not totally shake the trauma of being frozen. My appetite was gone; food no longer tasted good, and I had to force myself to eat.

Night blindness had returned and dysentery had shown up among the men, which is not only highly contagious, it is deadly if not treated properly and quickly.

I had developed diarrhea and feared it would develop into full-blown dysentery. My fears were soon to be realized. Blood and mucus began to flow freely with each trip to the outhouse.

The Chinese had issued small boxes to be used as sample boxes. Each man would send one. When it was found that a POW tested positive, the company commander was notified to transport that POW to the aid station to isolate him from the other POWs. The commander had no choice but to comply with the doctor's request. I had requested a sample box and was denied once again.

One morning Lieutenant Wu came along and saw a large puddle of blood on the ground and inquired as to whom that blood belonged. Cooper, a fellow POW, said, "I don't know, but Abbott was just out there. It must be his."

Lieutenant Wu said, "No, no, no, not Abbott."

And Cooper said, "why not? He was just out there."

Lieutenant Wu said, "That blood is red. Abbott's blood is red, white and blue," and he walked off.

It became clear to me that the company commander was not going to allow me to send my stool sample to the lab. It was also clear to me that if I did not get treatment I would soon join those buried up on the hill.

I found a box that had been given to another man who was healthy, and once again wrote his name on the top of the lid and my name on the bottom.

The usual practice was for men going to the lab to have their stool sampled would carry other samples along. When this man got to the lab, he turned the lid over and exposed my name, and said the doctor took one look at the sample and spoke to the guard, who headed out to the company to get me.

When the guard went to the company commander and told him I had dysentery and was to be taken away and placed in quarantine, he blew his stack. "No, no, no!" But the guard insisted.

The commander called me to his office and gave me specific instructions as to what to do and how to act while at the aid station. I was not to leave the building or associate with anyone, and if caught out in the street socializing I would be turned over to the Koreans for punishment.

The treatment for dysentery was a very simple one: you would get two large glasses of Epsom salts and drink two gallons of water, and eat all the noodles you could hold three times a day. It worked.

The Chinese cook in charge of making noodles for the entire camp was my teacher. I worked with him in the early days at Camp 1, and he taught me how to make noodles unassisted.

The cook inquired as to my health status. When I told him I was recovering and going back to the company soon, he asked if I would stay on and cook for the sick. I said I would. I was not surprised at his request because I had become quite good at making noodles.

I soon discovered a small shed out back that was used as a meat storage building. There were hogs quartered and hanging out there. I was determined to get some of that meat for those poor devils that were recovering from dysentery. My problem was getting the meat without the Chinese knowing it.

The Chinese used the meat for seasoning and would cut the meat up into very small pieces, but I had visions of sliced ham every day.

The Chinese cook took inventory of the meat house and I decided the best thing for me was to slice a little bit off of each ham and cook it with the rice, which was cooked every morning. Since I could not fry the ham without being caught, I placed the ham in the bottom of the pot and cooked it along with the rice.

When I would steam the rice, there was a heavy crust left at the bottom of the pot and the ham was embedded in the crust. I soon found I could season the rice and cut the crust separately without the Chinese knowing it. When I cut the crust and meat up together and fed the men, they could get nutrition from it. The men had no idea why that rice and crust tasted so good.

The threat of dysentery disappeared. The Chinese closed the aid station, and I went back to the company. The company commander thanked me for being such a good cook and helping with the sick.

The commander's attitude had changed. He had nothing to say to me after that. For about a month I noticed the guard at each shift would come into my room and shine a light into my eyes and walk away.

I remembered Wu telling me how they would do that to him and his family before coming to Korea. I told Pat Patterson that I would soon be leaving and cautioned him to be careful, keep his mouth shut, work hard, and he would go home one day. But Pat did not return home as far as I can find out.

One morning the Chinese cook supervisor came to me with an interpreter to ask if I would help him make noodles for the camp. He said, "You're Number One."

I thanked him and said I would gladly do it, provided I could move out of the company and away from the company commander. He said he had no authority to do that, he was just trying to help me, and he walked off.

I knew then my suspicions were correct. I would be leaving the camp soon. I had no idea where I would be going, but judging from my friend's look, it would not be a pleasant place.

That night, around twelve o'clock, the guards came and got me and told me to take all of my stuff. We walked about a half-mile beyond the camp. As we went around the bend in the road, I could see two trucks with men standing around. My first thought was that they were Russians because they were much larger than the Chinese. I later found they were actually British and American POWs.

As I joined the others, John Kirby and the others from his squad were standing together. His group had always been separated from the general population and were not allowed to speak or associate with the general prison population. I had been placed in that squad when I was doing hard labor and was told by the guard as he pointed toward China that they would go to a peace camp.

John said, "What do you think?"

I said, “I don’t have any idea, but I’m sure of one thing, John. They will not shoot us.” I said, “Over there stands Bates.”

That was the same Bates that the guard had pointed out to me when I was cutting wood for the Chinese kitchen. He was the one announcing over the PA system, “End of conversation, no talking.”

Chapter 11
Officers' Camp and Company B

Twenty-one of us were loaded onto trucks and traveled for two days. We stopped and went through the officers' camp and then continued over the mountain where we established Company B.

The Chinese began to treat us better. We had better food than ever before. We were informed we were soldiers under oath and were going to be treated like soldiers. We were expected to work hard at building the facilities at Company B. We had wood detail, and we carried rocks from the river for the foundation of two basketball courts and a softball field. We were told that since we had sabotaged their indoctrination we would no longer be required to study communism.

There was only one escape attempt. Joe Adams tried to escape and
was shot in the leg. In later years, when he applied for the Purple Heart, Joe was initially denied because the war was over when he was shot and he did not know it. Eventually, however, he did receive the award.

We built the basketball courts and playgrounds, and when I think about that group of brave men, the old heart beats a little faster. These men were the best of the best. There were Englishmen, Frenchmen, and an Irishman among us.

The amazing thing about the English, regardless of the situation, is that they can always find some means to entertain themselves, including dramatic productions. Those in labor camp Company B were no exception. One of the Englishmen and the Frenchman went to the Chinese company commander and asked for permission to put on a play. The English would write the script, recruit the actors, put the whole thing together, and I thought they did quite well under the circumstances. At the direction of the Englishmen and the Frenchman, we did an outstanding job. I, of course, played whatever role I could—singing, dancing, it didn't matter to me.

The Chinese gave us two guitars and one of the POWs made another one. We formed a band to provide music for the plays. The Frenchman was an outstanding singer. We had a sergeant who played the spoons in the band and another man who played guitar very well. He came from eastern Kentucky where playing music is a way of life. His older brothers, known as the Shelton Brothers, had their music played over the radio. I did not know his older brothers, but the Shelton I knew was a credit to his family, country, and fellow man, and as good a man as ever "dropped a load between two hocks"—to use an east Kentucky expression.

In Company B, every man carried his own weight without complaint. Medical supplies were scarce, and dental treatment was non-existent. We had one man with a severe tooth infection. Several of the other prisoners actually pulled out his teeth, with no pain medication.

For the sake of propaganda, the Chinese established some "Olympic games" between Company B and the officers' camp. The Chinese officers wanted us to do well at the games. They selected the men they wanted to be in the games and allowed us time to train. We had a basketball team, a softball team and a track team.

When it was time to go to the Olympics, we loaded onto trucks to go over the mountain to the officers' camp where the Olympics were going to be held.

On the way we passed a South Korean POW camp, and there was a disabled ox cart out in the middle of the road. We had to come to a stop because the road was blocked. Suddenly I saw the young South Korean who had cut his foot off in Death Valley. He hobbled out toward the truck to greet me.

"Syng," I said, "is that you?"

He said, "Yes!"

I said, "You cheated me, you bastard! You've got too much fire in your belly to die!"

He laughed as he reached up to shake my hand, and then he fell down on his hands and knees. We were so glad to see each other. As we drove away I could see him crying, with tears flowing down his face.

We also passed a women's camp which had mostly Turkish women prisoners. It was thought they had been captured at the fall of the embassy in Seoul. There seemed to be cotton-headed young'uns all over the place, which was exciting in itself.

We unloaded the trucks when we arrived at the officers' camp, and entered a house. There was an old woman curled up in a ball on a rug in a corner. She didn't want to lose her house, but the Chinese just took hold of the rug and dragged her out. Once outside they pulled her to her feet and began walking her across the field to another village. She began crying and screaming, and one of the guards took her by the shoulders and shook her until she quieted down. We never saw her again.

We spent about four days in the officers' camp during the Olympic games. We did well in every event but we did not win any event—except one—but we never lost badly, either. We lost softball two to zero, basketball by ten points. Boxing, we won. We had two boxers—Matt Maddingly, light-heavyweight, and McNabb, heavyweight. In track, I ran in the five-hundred-yard dash. I led the race up to about fifty yards from the finish line when my left leg collapsed and I could not finish the race.

Our theater performances were successful. Our band worked out well. My left leg really hindered my dance routine. We wanted to do well, but it was the comradeship, the chance to meet and talk with our officers that was the high point of the event for me.

Major McLaughlin was on the officers' basketball team. I had great respect for Major McLaughlin as an officer and gentleman. I also enjoyed poking fun with him and found I could easily get under his skin. He was a Marine Corps officer of the finest caliber and such a competitor, very no-nonsense.

I remember Major McLaughlin on the march north when one of the Marine Corps officers lost it. He began complaining extensively. After several hours of listening to the officer complain and cry like a baby, Major McLaughlin walked over to the officer and reminded him that he was a Marine Corps officer, and that if he continued to complain and act like a child, he was going to knock the hell out of him. The officer never complained again.

While we were at the officers' camp I had wanted desperately to get an audience with Major Alley, but he did not want to talk to me. Major Alley was important to me because I had promised him I would see him court-martialed for his actions back in Death Valley, and I wondered if I had judged him too harshly. I wanted to know his true motivation.

I asked a lieutenant about Major's Alley's actions as a POW. He gave few good comments about him. In fact, I found he was, for the most part, looked upon with suspicion by his fellow officers, and I concluded that I had made the right decision. I did see him court-martialed, and I was a government witness at his trial. He was found guilty and sent to Ft. Leavenworth. After his release from prison he died a broken man.

When we returned to Company B, we learned that while we were gone, François had gotten into trouble for making sarcastic remarks about a young schoolteacher. That was the only unusual instance that took place while we were away.

The Chinese had stacked about a hundred cans of sake in front of a building. They were evidently getting ready for some kind of a celebration. They had brought five sows. We had the job of marching five miles down to the river to get ten more that had been brought over by boat from China.

I noticed that the guards would shake the cans of sake every night to make sure they had not been tampered with. I also noticed that they never moved the cans, so I decided to steal one of them. I removed one of the boards in the floor of a latrine and hid my can of sake there.

I had a pint jar that I sterilized and used to dip out the sake and would drink it at night in the room. Since it looked like water and had little smell, no one knew it was sake. One fellow POW did observe that I drank an awful lot of water.

One day I took my jar of sake and went down under the hill where they let me run to train for the Olympics. There was a large overhanging rock with the sun shining directly under it. I crawled under the rock and sat there enjoying the warm sunshine and my sake, too much, too fast. I fell asleep and woke up with the mountainside covered with Chinese soldiers. I slipped out from under the rock and ran back up to the room, and asked a man named Seaco what was going on. He said the Chinese were looking for me. I ran back in the room and put a blanket over my head and lay down in a dark corner, and asked him to notify the Chinese that I had been found sleeping in the room. The Chinese came to the room, saw me there, and called of the hunt without further incident.

The day after our return to Company B, several of us went down to the river to unload hogs brought over from China by boat, to take them five or six miles back to camp, and then butcher them. When returning with our hogs, each man was responsible for one hog. He tied a rope around the hog's leg and then walked him back to camp.

As we were walking down to the river to unload the hogs I heard a voice up on the hill saying, "No, no, no," in English. I could not help but wonder what was going on up on that hill.

On our way back as we approached that area where I heard the cry in English, I "accidentally" headed my hog up over the bank and up the hill. The hog ran, and I caught him near an open fence.

The fence surrounded a house where there were four Chinese guards, and four or five other men with white skin who were dressed in Chinese uniforms. I considered them to be Russians. There was one American officer standing near the well.

He was wet. I could see the water on his flak jacket.

I knew I had walked upon something I was not supposed to witness.

Just as I got my hog headed back down the hill, the American officer, who was a captain, broke away and ran up to me at the fence, and said "These bastards are putting me in and out of this well. Please don't believe a damned word I say or write."

I kept walking, and when I looked back, they had the captain on the ground.

One of the Russians yelled at me, “Hey you, get back here!”

The hog had decided it was going to join the others on the road and was pulling me in that direction. I ran as fast as the hog would let me run and I caught up with the others. I kept looking over my shoulder for fear that the group of Russians would follow me, but they did not. I never saw or heard of the captain again, except later a Chinese guard told me that they had captured an American officer and he had confessed to bug warfare in order to spread diseases.

The Chinese were so fearful of bug warfare they had required each POW at Camp One to catch five or more flies every day.

The day after we brought the hogs back from the river, we butchered about eight of them and the Chinese salted and hung the meat.

This added to my suspicion that the Chinese were planning a celebration.

The following day the Chinese commander assembled the troops and announced that the sick were going to be repatriated and would leave first, and the rest of us would be going to the 38th parallel to be released. He said “I am proud to have known you. We have captured many enemy troops during this war, but we have captured only twenty-one soldiers, and they stand before me this morning."

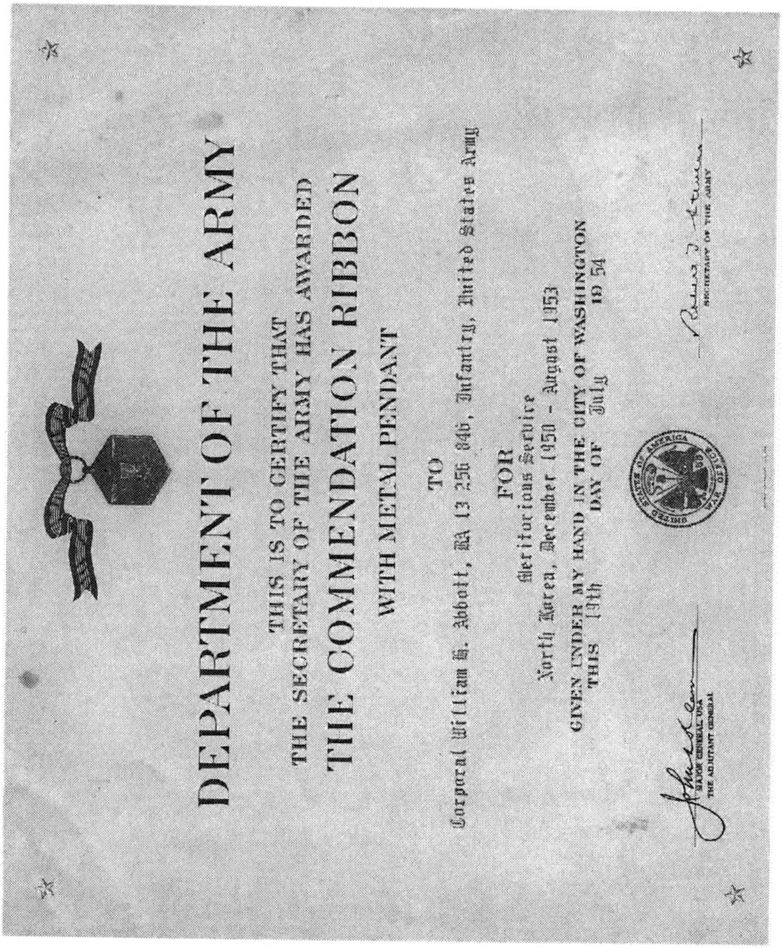
DEPARTMENT OF THE ARMY

THIS IS TO CERTIFY THAT
THE SECRETARY OF THE ARMY HAS AWARDED
THE COMMENDATION RIBBON
WITH METAL PENDANT

TO

Corporal William H. Abbott, RA 13 256 846, Infantry, United States Army

FOR

Meritorious Service
North Korea, December 1950 - August 1953

GIVEN UNDER MY HAND IN THE CITY OF WASHINGTON
THIS 19th DAY OF July 19 54

MAJOR GENERAL USA
THE ADJUTANT GENERAL

SECRETARY OF THE ARMY

Certificate of Commendation Ribbon with Metal Pendant

At that time, only twenty-two soldiers had received this commendation

Chapter 12
Back at Home

After the war was over we were moved from Labor Camp B to Camp 5, where we spent a week. Then the Chinese loaded us onto barges going down the Yalu River.

On our trip down the river we stopped to pick up twelve POWs who had elected to stay with the Chinese rather than return home to the United States. I knew some of these men and I expressed my disappoint-ment in them.

These men were, in my judgment, throwing away every opportunity to live a reasonably good life back in the states. One thing was obvious to me. These men, almost without exception, felt ashamed and were afraid to face the court-martials that were waiting for them back home. I tried to convey this to them as best I could.

"You cannot hide forever," I said, and the boat turned to the Chinese side of the river. The men began stepping off the boat onto Chinese soil.

I continued to express my opinion that they were giving up their last opportunity to come home. I told them, "Make no mistake about it. The Chinese don't like you. They will use you in one of their factories, and you will rot away. It's that simple."

The boat turned and headed down the river on the Korean side, where we got off and were loaded onto trucks headed to the railhead. There we were loaded onto a train and taken to Pyongyang, and from there we went to Panmunjeom, where the prisoner exchange was to take place. The exchange rate was eight Koreans to one GI.

An American second lieutenant had a roster set up on a small card table, verifying every American POW as the exchange took place.

For the first three days I walked down the hill and waited to be exchanged (since, after all, my name was at the head of the alphabet); but every day I was sent back up the hill.

On the evening of the third day I stood looking over the hill and wondered why my name had not been on the repatriation list. "The Chinese have not turned over my name; that has to be the answer!" I thought.

Suddenly I realized there were two Chinese interpreters standing behind me, speaking English. I glanced over my shoulder and saw that one Chinese officer was pointing his finger at me, and he said, "This man is Abbott. If he is not exchanged tomorrow, you will take him back north with you. He will join the captain and the eight other men you're scheduled to take back." They then walked away, speaking Chinese once again.

The thought of being returned north, perhaps to go to the Russian mines, was a very disturbing thought to me. My mind wandered back to the trip coming south.

The Chinese had done a tremendous job of repairing the train tracks after they had been bombed out by our Air Force, but the rails in most places were just temporary, and I did not want to ride the train. The train coming south had only been able to run at five miles per hour. I sat on the side of the boxcar watching the rails bend and sway as the train rolled across them. That was an experience I'll never forget: The rails would literally drop twelve to fourteen inches and the train would just creep across. It looked like a rubber band. But with luck, the train rolled on through.

At the repatriation center, the Red Cross was handing out cigarettes, toothpaste, shaving cream and mess kit equipment. Small things, like brushing your teeth, really brought back memories of how sweet life is in America.

Then I came back to reality. I was in Panmunjeom. I was to be repatriated but nobody seemed to want me. Why would the Chinese tell me I would be taken back north with the others? The fact was that the Americans did not want me, and the Chinese were not overjoyed that I might be staying with them. I thought, "Damn, Abbott! You're some kind of bad dude. Not even the enemy wants you!"

The next morning as I started back down the hill, a good POW friend of mine wished me luck. I said, "Thank you, John, but I won't be coming back up here, regardless of how it falls."

On my way down the hill, I kept thinking, "Since the Chinese don't want to keep me, then they won't prevent me from working my way up to the front. And they won't stop me from going through the gate."

I worked my way up to the front of the line. My suspicion was correct. No one was complaining except the GIs, who were saying, "The line begins in the rear."

They didn't realize that it didn't matter where you were. If your name came up in alphabetical order, they were going to take you. I ignored their remarks and just kept going. I ended up in the third position in the front of the line. The first two men went through the exchange point without a hitch. The lieutenant went through his book to find my name as he had two days before. This time, however, I didn't wait for the lieutenant to tell me I was not on his list. The guard unlocked the gate, and I kicked it open, knocking the lieutenant, his table, and his book over the hill. I then ran through the gates to safety.

The lieutenant was hollering, "Stop that man!"

I ran directly into the arms of a full colonel. He said, "What's going on here?" I told him I was not on the list, and to speak to that second lieutenant.

As I looked at the flags flying there, I saw no American flag flying; instead I saw only the UN and North Korean flags.

I said, "Colonel, after all of the men who have been lost in this war, one would think the United States could find an American flag."

He said, "They did not brainwash you, did they, Son?"

"Colonel, it was very simple. They couldn't find anything to wash."

The government had set up a special area for processing the POWs. The nurses handed me a questionnaire. I answered all the questions to indicate that I had no medical problems.

It was a lie, of course. I was sick as a dog. I could not eat and had been coughing up blood for about six months. Now the blood had stopped and I had only a dry cough and a sore chest. My shoulder ached and I had very limited use of my right arm. My legs and feet hurt. The pain in my back and shoulders had been with me since the Chosin reservoir. I thought surely it was not life threatening any longer.

The only thing I had in my mind at this point was, "I must get out of here and away from the Chinese." I tried to get as far away from the Chinese as I could, as soon as possible. My study of Communism had taught me not to trust people.

I could not get over the fact that I was not on the repatriation list, that the enemy did not want me, and that the American government did not want me.

I left the nurse's tent, walked to the helicopter landing-pad, answered to someone else's name, and got on the helicopter.

We flew to Inchon where they issued new uniforms. I did not report for duty, but instead found and boarded the ship that was sailing to San Francisco.

When the ship was well underway, I reported to Intelligence and learned that I had been reported as killed in action.

The search was on to ascertain my identity.

I made a report on two POWs who were involved in collaboration with the enemy. One of the soldiers was Major Alley, and the other was Private Bates. Both of these men were later court-martialed. Major Alley was found guilty, and Private Bates was found not guilty. I did not testify in the Bates trial.

We landed at the Navy Base in San Francisco. After chow, we went to the payroll office where I was advised that I could get as much as $5,000. I had the payroll clerk give me the entire amount.

I had a thirty-day leave before I had to report for duty at Fort Meade, Maryland. I got a money order and sent $4,800 to my mother and bought an airplane ticket to Wheeling, West Virginia with the balance.

I was picked up at the airport by my oldest brother, who told me of the problems my parents had had with the Army. My father's account of the government's actions after I was reported killed in action were something of interest.

When I was hit the last time at the Chosin reservoir, the medic had picked up my dog tags and turned them in to the Army. The Army reported me as killed in action, paid off my insurance benefits and all money I had coming to me. They would want all that money back, as I would learn later.

Of course, I was not aware of what had taken place until I got back home. My father told me that when he received the first check, death certificate, and Purple Heart from Washington, he called the Pentagon and told them of the note he had received from me that had been delivered to him by a POW named Larson who had been released in 1951. He was from West Virginia, just across the river.

The Army would not accept any proof that I was alive. They had my dog tags and were unwilling to admit I was alive. They believed that they, in fact, had sufficient evidence to prove that I was killed in action on December 1, 1950.

The Army just did not want to hear about it anymore. Their remark to my father was, "Cash that check so we can keep our records straight. We want to close the chapter on this soldier."

A few months later, a call came from the Army asking if my father wanted my body brought back home for burial. He told them "No," and once again explained, "My son is not dead. One day he will come marching home, alive and well, I trust."

My father told me, "I did not hear from the Army again. I was watching TV when the POWs were being unloaded at San Francisco and saw you on the gangplank as it was being dropped down. The radio then announced the POWs' names. The Army had made a big mistake, Son. They would not listen to reason. I hope that don't get you into a lot of trouble."

I told him, "The Army will realize their mistake, Dad, and that will be the end of it."

That did not prove to be the case. The Army wanted its money back, including the $5,000 they had paid me in San Francisco.

While I was home on leave, I took my youngest sister to a basketball game at the high school where she was the coach, and I met Betty, the home economics teacher. We went to dinner, and began to date. Six months later we were married.

When I reported back for duty at Fort Meade, I did not have a complete uniform. When I went to get a uniform, the supply sergeant informed me that I must pay for it.

Having to pay for my uniform made no sense to me. When we left Japan for Korea we turned in two duffle bags full of good uniforms for storage, which apparently had been lost by the Army. Now they expected me to buy uniforms.

"To hell with this!" I said. "If the Army cannot or will not provide me with uniforms and clothing, I'll not stay, Gentlemen."

Then I went to payroll, and found they withheld most of my pay. I was given only ten dollars of it. The severe reduction in pay was to continue monthly until all moneys were paid back. How would I ever be able to pay that amount of money back?

The payroll clerk told me that the Army had mistakenly advanced me the $5,000 when we landed in San Francisco. Luckily I had wired that money home and had not deposited it in the bank where the government could tie it up and freeze the account, or I would have spent all that time in the POW camp without any compensation at all.

That $5,000 was gone in a very short time. At least, I thought, my time as a POW was not all for nothing.

I don't think it was the money the Army wanted back that upset me as much as it was them refusing to furnish me a uniform. That was the last straw as far as I was concerned. They could all go to hell, and I would go back to milking cows and growing corn.

The following morning a company clerk knocked on my door and said the platoon sergeant would not be there until later in the morning, and told me the first sergeant wanted me to fall the platoon out for reveille. I told him I could not do it because I didn't have a complete uniform, and that I would see the first sergeant a little later in the day. I went downstairs and asked the senior squad leader to handle the reveille.

I went to the company commander's office and reported to the first sergeant. He was a crusty old top sergeant. His first words to me were, "I'll have your stripes, Abbott." This was because I had not handled reveille.

"That will be no problem, Sergeant. I'll gladly give you these stripes. I've come here to see the company commander."

I told the commander I did not have a uniform and appealed to him for clothing and permission to check into the hospital because I was still coughing up blood and could not eat.

He sent me to sick call, and a corporal put a bandage on the wound on my back, which was seeping a little bit. I wanted to see the doctor, but the corporal would not allow me to do that, and he did not record in my file the fact that I was coughing up blood.

I was sure as hell not going to have any wet-nosed orderly determine how my wounds were to be treated.

I went to the hospital emergency room. There was a staff sergeant on duty there. He informed me there were no doctors available and I would have to come back in the morning, but I did not go back.

The next morning I went back to the company commander and the first sergeant said, “The company commander does not have time for you.”

The company commander’s door was open and I could see the him sitting behind his desk.

“Is that not the company commander sitting in there?” I asked.

"Yes," the sergeant replied, "but he does not have time to see you, Abbott."

"Well, Sergeant, what damned good is a company commander who can't take the time to see his troops?"

I walked into the commander's office. He had heard the conversation between the first sergeant and me. He was like a tiger ready for his prey. He jumped out of his chair without returning my salute. Both fists came down on the desk. You could hear him yelling, and his response shook the room.

"Just hold on a damned minute, Captain," I said. "I just came here to inform you that I am three years overdue for discharge, and before you spoil your day, Sir, here is my telephone number. Call me when my discharge papers are ready," and I walked out.

I looked at the first sergeant, and said "Top, when you see me walk out that door, you're looking at the back side of the best damned sergeant that ever walked through it."

I heard him tell the captain, "There goes one crazy son of a bitch."

I left the base and drove to my sister's home in Washington, D.C.

Thirty days later, when my discharge was ready, I went out to Fort Meade to pick it up.

I was not given a discharge medical examination. There was just a formal run-through. I explained to the major processing me out that I was as sick as a dog, and he said, "You're out of the Army now. That's your problem."

An old colonel was standing outside and he had overheard the conversation between the major and me. The colonel assured the major that if he pushed one more POW out the door without medical documentation, he would be passed over for promotion. He then turned to me and said, "The only thing I can do for you now, Son, is to write a letter to the Veterans Administration on your behalf."

I said, "Okay," and walked away.

I thought it sure would have been nice had I met this colonel a bit sooner. Things probably would have been a whole lot better.

I hated leaving the Army on such a sour note, but I still had that terrible feeling that the Army would just as soon get rid of all the POWs and hope they would just disappear, never to resurface again.

I felt that had the circumstances been different, I would have re-enlisted, or at least extended until I could get medical treatment.

Had my pay not been frozen, it would have made every bit of difference. I could not survive on ten dollars a month, nor would I try. The issue of buying uniforms when the Army had lost mine was a no-brainer for me. I would not put up with that. I needed medical attention and would take my chances with the VA. That also turned out to be a terrible mistake.

I thought that once I was out of the Army the government would stop hounding me for the money they had mistakenly given me, but they continued pressuring me to return it.

I moved back to live with my parents on the family farm in Ohio. When I got home I found that I could not handle the bales of hay. I decided I could not do farm work any longer. The pain in my upper back and right shoulder was unbelievable. Lifting anything, particularly over my head, was literally impossible. I was living on milk and crackers because my digestive system would not tolerate solid food, and my feet were so swollen I could not wear normal shoes.

I reported to the VA for a physical examination. They were not sympathetic to my cause, as to the service connection of my injuries. In fact, the doctor asked if I was a hunter.

"Of course I was a hunter as a boy," I told him.

"Well, where were you hunting when the accident occurred?" he asked.

"What accident?" I responded.

"The one when you got shot in the back."

I became furious, and walked out. He marked my records, "Veteran is uncooperative."

I did receive a ten percent disability compensation for what is now termed post-traumatic stress syndrome. It was just called a nervous condition back then.

About six months after my discharge I received orders to report to Fort Meade to be decorated for my actions while in captivity. On July 19, 1954 during a dress parade at Fort Meade, Lieutenant General Floyd Parks pinned the citation on my uniform.

General Parks had been notified of my financial plight; that the Army wanted to recoup all their money, and he assured me, in his words, "Go on home, Son, and don't worry about it. You won't hear from them about it anymore."

On our way off the base, we stopped at the gate and I was informed that the colonel who was General Parks' aide wanted me to join him for a drink in the officers' club.

When we met, he said, "The general wants you and he has an important job for you. He told me about a new anti-communist school that he wants you to participate in."

The colonel advised me that I would be coming back into the Army as a senior warrant officer (S-4). He would secure housing for me and my wife, who was pregnant with our first child. Betty would have a job teaching school on the post after the baby was born.

I turned him down because I was totally burned out. My remark to the colonel was, "I don't have an ounce of soldier left in me. I would be a disappointment to my country, my family, and General Parks."

The colonel asked me to see him the next day at the recruiting area but I did not show up. Later on he sent me pictures of the decorating ceremony but I never responded to his offer.

I went back to Ohio.

About two months after I was decorated, I received orders to report to Fort Meade for the court-martial of Major Ron Alley. Mrs. Alley invited me to dinner but I was advised by the government prosecutor not to accept her invitation so I declined.

At the court-martial I spoke off the record and suggested that, in all fairness, it might be wise for the Army to take a second look and perhaps make a determination as to an officer's combat qualifications, not just as an administrative officer.

I also made the point that we were facing the most dedicated enemy the nation has ever faced. I said, "This enemy, communism, will destroy every fiber of our nation, drain us physically and financially and destroy us internally. The Communists will destroy us not with guns on the battlefield, which we are ably equipped to handle, but they will plant the seeds of propaganda in the minds of freedom-loving people of a utopian world; a propaganda war that is far more deadly than the shooting war that we are accustomed to. America will have to adjust to that threat.

"In all great nations, history tells us that the final diplomacy of its nation is the military. After all the diplomatic types have gone, the troops are all that remain. In short, every man, regardless of rank, must be equipped to meet this war of ideas.

"Who is to know what's in a man's heart? But ways must be found to determine what is the mindset of those in uniform as it relates to his patriotism, love of country, and commitment to our form of government and our way of life.

"As for Major Alley, I leave the presumption of guilt or innocence, right or wrong, to this learned body of fair and just men sitting here today."

I took a job as a service repairman with General Electric Distributors and moved from Cincinnati, Ohio to Huntington, West Virginia. I attended night school and studied electricity, heating, air conditioning and refrigeration.

When I went to the VA in Huntington, they admitted me to the hospital and ran a battery of tests on me. They did not treat my digestive system or my wounds, but they did give me extra heavy socks to protect my feet, which seemed to help a little bit, and advised me to move to a warmer climate.

The doctor predicted that within five years I would lose the use of my right shoulder and arm, but no additional service-connected compensation was granted. Under the Eisenhower administration, unless there were absolute medical records established through the military and there was a record of treatment for those wounds, there would be no "service connection," regardless of the nature of the wounds. I could get a hearing before the compensation rating board, but I had to get other POWs to substantiate the cause of my wounds. They would also have to verify the conditions existing in the POW camp.

I called Kirby, an ex-POW and a dear friend. He wrote a letter verifying the circumstances surrounding the injuries to my right shoulder and the problems with my back. I finally got "service connected" status for it, but only after a long battle.

About a year later, I moved to Daytona Beach, Florida with a contract for a new job with Goodyear stores and left my family in West Virginia. After about three months I took my new truck to Huntington to move my family and our belongings back to Florida.

While living in Daytona Beach I attended night classes and this time I studied electrical engineering.

I saw an employment ad in the newspaper for an industrial electrician at Cape Canaveral. I applied for the job and was hired immediately. I commuted from Daytona Beach for several months until my son was old enough to move away from his pediatrician. Then we moved to Titusville and settled down. I worked at Cape Canaveral as a high voltage electrician for about a year. One day I lost the strength in my right hand and dropped a pair of pliers into a piece of high voltage equipment. If I had dropped it a few inches in any direction other than where it fell, the equipment would have exploded and several men would have been killed or at least badly burned, including me. I walked to the personnel office and turned in my resignation.

Three days later I saw Dr. Akers, an orthopedic surgeon in Titusville, who determined that my shoulder had been driven down in excess of one inch. The rotator cuff had been crushed and was blocking the nerves.

He sent me to Bay Pines VA Hospital in St. Petersburg for surgery. Dr. Barlow, the chief VA surgeon, operated on my shoulder, and the next morning as the doctors made their rounds, he remarked to the other doctors, "I feel sorry for this one. He will never use that arm again."

After the operation I could feel the movement in my fingers which I could not feel before I had the surgery. I motioned for Doc Barlow to come a little closer, and I whispered in his ear as I showed him the movement in my fingers, "Doc, you're full of shit; look." He saw my fingers move, and laughed.

After about two weeks of my shoulder feeling fine in traction, I was awakened by severe pain in my right shoulder. There were no nurses stationed nearby, and no bell to ring the nurse's station. I unhooked my traction bars and walked back to the nurse's station. There was only one male nurse on duty.

I told the nurse, "Did you know that I had surgery two weeks ago and my bandage has not been changed since then? Would you please change my bandage?"

He said he was not authorized to change bandages, and that the head nurse would have to do that.

I told him that I had developed extensive pain in my shoulder. "I fear I may be getting an infection."

His reply was "If you want it changed, you can change it yourself. The head nurse will change it for you when she gets back on Monday."

"This is Saturday," I said, "I can't emphasize too strongly that there's a lot of pain here. I need this bandage changed. Who do I see to get it changed if you're not authorized to do it?"

He was a smart-ass, and said, "I repeat, if you want the bandage changed, you will have to change it yourself." I started pulling at the tape, as he sat there and watched with his hands folded.

I finally pulled the tape down to the incision, and when I pulled the bandage off, the smell was terrible. It was enough to make you sick. The nurse said, "Oh my God!" He ran over and called Dr. Barlow, who was actually just down the hall. He told the nurse to call Dr. Brown, that I would have to go directly to surgery first thing in the morning.

At seven a.m. I was taken to surgery. The first words out of Dr. Barlow's mouth were, "Son, you'll be the luckiest man who ever lived if you don't have osteomyelitis in that shoulder."

The whole cause of this episode actually happened when they prepared me for surgery two weeks earlier. In those days, I had a lot of hair on my chest. They shaved my chest and left the sponge used to wipe away the hair on the operating table. The nurse inadvertently picked up the same sponge and wiped my wound with it. The sponge filled my shoulder wound with the hair from my chest, causing an infection.

It was eighteen months before the cast was removed and I was able to use my right arm at all. While I was in the hospital, the VA had given me a hundred percent service-connected rating which paid the basic bills, but it ran out three months after my surgery.

It was sad. For first time since leaving night school things had been looking great. My brother was helping me run our contracting business in Daytona Beach and it was going well, work at the Cape had been going well, I had money in the bank with all my equipment paid for, and a new family car paid for. We had planned to build a new house. All that was now to change because the infection had prolonged the healing of my shoulder and I could not work with my right arm.

By now, Betty and I had three children. The baby was six months old. Betty and I sat down at the dining room table and asked ourselves, "What will we do now?" As we looked to the future, it seemed very dark.

We decided to close the electro-mechanical business we had, sell the trucks and equipment, and Betty would go to work. She was a dietician. Our neighbor, Miss Barker, would keep the children.

We went down to the river with the children and went fishing. Betty had not said two words since leaving the house. I walked over to her, put my good arm around her shoulder, and assured her that everything was going to be all right.

The children were playing in the sand and I said, "Let us thank God for all the blessings He has given us. He has given us three beautiful children, and the means by which to feed and clothe them." I said, "Think about it, Bet; even a chicken with a broken wing can still scratch for her chicks. Can I do less? Should I do less? I think not. Hell, I'll have this thing worked out by morning, with God's help."

That evening a friend of mine named Dan Sidetack stopped by. He was in the insurance business and was doing very well. I was in the insurance business the very next day. I went to work for him and started studying for the state exam. I passed the exam, went to Orlando College in Orlando studying insurance, opened my own office, and never looked back.

Over the next two years my business expanded to included real estate, mortgages and mortgage banking, and class-one building contractors.

I met a man I had known when I was in the electrical contracting business and he offered me the opportunity to return to the Cape three days a week. The direct contact with so many people at the Cape allowed me to write a lot of auto insurance that I would not have been able to do under other circumstances.

At a special dinner held at the Cape one evening, I met the director of vocational education at the local junior college, who was planning a new appliance refrigeration course. I told him I would be interested in teaching the course. He offered me the position and I taught electricity, power distribution, and appliance repair for several years.

I completed my vocational teaching certification from Florida State University, and was tenured at the college and served on the College senate.

With my family secure and my finances in place, Betty and I became involved in politics and decided to help build the Republican party in Florida because, at the time of my registration as a Republican, there were not enough registered

Republican voters in Brevard County to even have a primary election.

Betty died of a massive heart attack at the age of fifty-eight and I remarried a year later.

Now, in my twilight years, I still enjoy the hustle and bustle of politics, managing my business affairs, and writing books.

Epilogue

On September 1, 2001, after some forty-five years of various delays during the work on this manuscript, it was finally complete and ready for printing. Then, on September 11, 2001, upon hearing of the terrorist attack on the United States, I decided to hold up on the printing of *Blood Runs Red White and Blue*.

As fate would have it, the following day I was involved in a severe car accident. The impact of the accident caused permanent loss of vision in one eye and five years of rehabilitation before my right hand was healed enough that I could begin writing again.

In January, 2007, I once again began the process of getting the book to press. However, there were delays due to hardware and software problems, and personnel issues. After overcoming these hurdles, on August 22nd 2007, my copy editor, Dolores, remarked that "the devil obviously does not want you to publish this book." On that note we stopped working and went to lunch to celebrate the completion of the final edit. As we were walking into the local Cracker Barrel restaurant I stumbled and fell and broke my hip in three places. After my surgery, while Dolores was visiting me at the hospital several days later she assured me, "The devil will not have his way. Your book will be published."

As I write this Epilogue I feel a great sense of relief, knowing that, as in the past, in the presence of God there is hope. This is being written from a wheelchair, but it is being written.

Every day I remind myself that God has a plan for each and every one of us, and for America. The whole world looks in good faith to America for help for food, shelter, protection, and guidance.

As I face the twilight years I realize the writing of this book has unraveled my very soul. The research and rekindling of memories of friends and events were in some cases refreshingly pleasant, but other past memories were unpleasant and probably would have been better off left alone.

In my mental rearview mirror I see this great nation falling apart, brick by brick, without qualified leadership dedicated to maintaining a free and independent nation called America.

From the 1950s forward I have seen our political climate change, redefining the terms "Communist" and "socialist" to "liberal," "leftist," or "radical." I have seen our federal government intrude into our everyday lives under the guise of "protection." I have seen America create "the great society" of socialized education, socialized medicine, and a welfare state. Our leaders and judges have diminished the roles of families, marriage, the rights of individuals and their freedoms, and they have worked relentlessly to remove morality, decency and religion from our free society. Individual freedom of speech has been superseded by political correctness for the good of the downtrodden few, changing our melting-pot society into a mixture of multiple opposing groups demanding that special treatment be given them both individually and collectively. History books have been rewritten and "sanitized" to reflect these changes in thinking. The Constitution—the document that guarantees our freedoms and is the basis of our laws—is now presented as "just as a suggestion."

In the past fifty years American society has followed the teachings of Karl Marx almost to the letter. Sadly, I see no change from this trend in the future. Many of those who are in seats of power, and some of those who are front-runners to be our next president, are separated from Karl Marx by time only. The U.S. Senate is set up to guarantee a continuance of socialistic policies for itself. A prime example is when New York elected a senator to represent the people's interests but found they had elected a full-fledged Communist fellow traveler whose goal is an anti-American, anti-Christian, pro-socialist leadership where third-world countries control all. Hilary Clinton, dedicated and enthusiastic Socialists, is well known for her socialist beliefs and activities, starting with the Vietnam era. She is a believer in the United Nations, whose socialist policies are directed and reflected by their socialist agenda. This Socialist/Communist agenda is plainly written in the United Nations charter.

All branches of our government are promoting the uprising of petty dictators. There is not a dime's worth of difference between those currently running for president. All are in lock-step, promoting a New World Order.

We have troops fighting for freedom. For whom are our troops dying in the Middle East? We have Congress and the courts setting up special minority classifications that divide all of America into groups.

In every part of America there are newspaper ads that promote minority set-asides, denying advancement opportunities to those who have lived and contributed to this nation all of their lives. When an American soldier leaves for the battlefield to protect us he leaves a job open for an illegal alien to cross the border and take it.

I noted the other day that with the classification of the troops returning from overseas, of the dead, seventy-four percent are of no minority classification. There is no set-aside for the brave men and women who made the ultimate sacrifice for our freedom. It seems that blood is all the New World Order wants from these noble individuals.

During my time as a POW I was indoctrinated in the Communist philosophy, and its plan for world domination is not one of confrontation, but of propaganda and guerilla warfare. The free world has been led to believe that today's terrorists are just a handful of extremist religious zealots, when in reality they represent a well-organized machine, well financed and truly zealous for the Communist goal of world domination.

The perfect policy for the Communists to exploit was the infamous Truman Doctrine, which puts forth a policy of containment instead of victory, not confronting Communism as the paper tiger it is but instead giving it credence to the point of equal power It has America announcing to the world that our policy is one of containment and not confrontation, thus allowing America to participate in wars but not win them.

The entire world looks at us in amazement and then exploits the opportunities provided by our way of life. The U.S. government and corporate America in are in bed with those who advocate the destruction of the capitalist system, with our freedom hanging in the balance.

Some way must be found to convince corporate America that it is in their own best interests to maintain allegiance to America and be good U.S. corporate citizens by research, development, manufacturing and selling their products in this country. There is nothing wrong with increased profits and stock price increases, but it is dangerous to use tax payers' money to complete research and development and then sell products to an overseas manufacturer, depriving our citizens the opportunity to have a decent standard of living, and at the same time expect the American government and our troops to bail them out to protect their interests. It is morally wrong to use American taxpayers' money to develop an American product and sell it on the international market at the expense of the taxpayer. Corporate America has allowed short-term profits to blind their vision. It makes no sense to develop a product, in many cases paid for by the taxpayer, manufacture that product overseas and then charge our citizens ten times more to purchase the product in the U.S. than they could purchase those products overseas.

American business has firmly hitched its future to the New World Order's economic wagon, which has our manufacturing base declining and American jobs drifting away, being outsourced to countries whose whole goal is the destruction of America and its biblical roots.

We continue to hear about the decline of Communism and that Russia is moving rapidly toward a free society. This dangerous thinking could not be further from the truth

Communism is stronger today than at any time in history. Today Communism is international in scope and it is wearing all different kinds and colors of hats. Romania, Poland, the Baltics, Russia, the

Muslim nations and China all embrace it. Even now Latin American Communism is on the march toward our open border. Chavez has nationalized Venezuela and is spreading Communism with financing and support from Iran.

The Bush administration and our Congress make no effort to close our borders to illegal invasion from around the world. American citizenship is free, without cost or obligation, to anyone who is oblivious to the sacrifices made by American men and women who have helped to establish and preserve our way of life. Our government is making every effort to accommodate their arrival with free medical care, free food, free housing, and free education, while accommodating their native languages and history, totally without regard for the sovereignty of this nation. Our government has set-asides (special treatment) for the minority status of these illegal aliens, according to the principles of Communism and the left.

I now see that the Communist influence is everywhere within the anti-Christian movement. To the American left (socialist) the end justifies the means. The left picks and chooses what laws to break and what to enforce; after all, laws are only suggestions, regardless of what damage it does to the victims or to America.

To the politically correct and the silver spoon crowd it's okay to steal and rob others if you are poor; it provides your children a better way of life. It's okay for illegals to butcher livestock along the border because they need food; no mention of the devastation and loss to the family's income that pays taxes and makes this country great. So they no longer count? It is evident that hard-working, tax-paying citizens are like sheep being led to the slaughter.

Thousands of people (especially the elderly) die every year, being executed after giving all their money and personal possessions to a cold-blooded robber who blows their heads off.

Congress passes laws giving a false sense of security to our citizens, but then the judicial system does not enforce the laws, or interprets the laws so that criminals have more rights than their victims.

I recall in 1953 just before the war ended I watched an old Korean woman being dragged from her home, kicking and screaming. I got on my knees at that moment and thanked God for the opportunity to live in a free America where no one can be forced to give up his life and worldly possessions without due process. Was my prayer premature? You be the judge.

Every day in America, some senior citizen is being forced from his or her home because some member of their family wants to take over her or his property and/or their money.

The quasi-police—self-appointed groups of citizens who, without investigation or verification of the facts—take it upon themselves to classify an older citizen as being abused and/or senile, and remove them from their home to a "protective" location, showing no regard for their rights or the rights of others. There are no avenues available with which to hold accountable these well-intended groups for the damage caused by their unsupported actions. This is a personal and true situation, and not an isolated incident nor a hypothetical case.

Some of today's young people sit like vultures waiting for their parents to die so they can take their belongings. I personally know of an eighty-four-year-old woman who was removed from her home, and her life savings pillaged by her adult children who did not allow her any contact with anyone outside of their small group, not even her husband, for two months, even though her husband was in the hospital. All the woman wanted was to visit her husband in the hospital and live at home in peace.

This woman's wishes were not to be met. The whole process was all done with out a search warrant or court order, and when brought to the local authorities' attention, they did nothing. My experience as a POW has given me the insight to understand their actions. The Communists used the same tactics in brain washing.

Could this have been done twenty or thirty years ago? I doubt it. I must ask the question: Have our individual freedoms eroded that much so as to allow this to happen in America? It is evident from this example, and many others, that they have.

I have seen many trials and tribulations in my life. My childhood during the great Depression influenced my commitment to God and my family, and provided me with my drive and blue-color work ethic. My time as a POW strengthened my love for my country, and the indoctrination of Communism strengthened my love for the freedoms and blessings of this great nation.

In the years since my internment as a POW I have been active in the American political system and have had a strong commitment to protect this nation and its citizens from the Communist way of life. It saddens me to see how far toward the Communist system we have drifted.

May God bless America, and may God bless us all.

Footnote

[1] William C. Bullitt, the first U.S. ambassador to the Soviet Union, never missed an opportunity to warn FDR of Stalin's treachery. But Roosevelt responded: "Bill, I don't dispute your facts; they are accurate. I don't dispute the logic of your reasoning. I just have a hunch that Stalin is not that kind of man. Harry [Hopkins] says he's not and that he doesn't want anything but security for his country, and I think that if I give him everything that I possibly can and ask nothing from him in return, *noblesse oblige*, he won't try to annex anything and will work for a world of democracy and peace." —William C. Bu`llitt, "How We Won the War and Lost the Peace," *Life* (August 30, 1948).

WHO'S WHO ON THE FAR LEFT
COMMUNIST PARTY–USA (CP-USA)

Position: U.S. branch of the International Communist Party. Policy comes directly from Moscow

Leadership: Gus Hall, general secretary; Henry W. Winston; Claude Lightfoot; Michael Zagarell; Daniel Rubin; Herbert Aptheker.

Location: Headquarters, New York City; membership estimated at between 12,000 and 13,000

Character: Predominantly white adults.

Brief History: Founded in Chicago, Illinois, September 1, 1919. Poverty and the U.S. depression helped strengthen the party during the 1930s. It polled 100,000 votes for its presidential candidate in Hitler-Stalin Pact of World War II, preceding the German invasion of Russia, caused many U.S. members to drop out of the party. After the war, CP-USA went underground to escape the 1940 Smith Act, which made it a crime to conspire, advocate, or teach the violent overthrow of the government; and the 1950 McCarren Act, which required registration of members. By the mid-1950s Senator Joseph McCarty had died; the Smith Act was gutted; and the McCarren Act registration requirement was ruled unconstitutional. The party returned to its open activities.

Although the Moscow brand of communism is considered reactionary by much of the New Left, party membership has increased about 25% since 1960. In 1968, the party had an official presidential candidate on the national ballot for the first time in 28 years. July 1968, Zagarell reported that "student unrest on college campuses and anti-draft demonstrations have been helped along by the Communist Party." The party's youth group, the W.E.B. DuBois Clubs, had less than a hundred members in March 1969, but party officials claim much of their own new membership consists of young people. At the May 1969 national convention, Hall cautioned, "it is not yet time to organize armed struggle;" but the party's Commission on Black Liberation approves "cooperation" with BPP.

Courtesy of "**Communism and the new left**": BOOKS by U.S. NEWS AND WORLD REPORT

1968 PRESIDENTIAL CONVENTION

Photo Courtesy of "**COMMUNISUM AND THE NEW LEFT**" BOOKS by U.S. NEWS AND WORLD REPORT

About the Author

WILLIAM H. ABBOTT was born in 1929 in Pricedale, Pennsylvania, the seventh of eleven children. He grew up during the era of the Great Depression. His father was a part-time farmer and a fulltime coal miner, working mostly in the coal mines of eastern Kentucky.

Mr. Abbott joined the U. S. Army and served with distinction as a combat soldier in the Korean War. He was captured and held as a prisoner of war for almost three years, during which he was reported as killed in action.

When he returned to civilian life, his back and shoulder combat wounds and the extreme cold and hunger of his days as a prisoner of war left him unable to do the work required by the family farm. Wanting to provide for his family, he knew he would have to find another means to make a living. He believed that education was the practical answer, without which he could not function.

Mr. Abbott spent ten years studying at colleges and universities, mostly attending night classes while working during the day. His occupations were extensive. He became an electrical and mechanical contractor, a class-one building contractor, a licensed insurance, real estate and mortgage broker, and a professor at a local college. He never worked a simple eight-hour day. He worked until the job was done.

He has based his life on the premise, "Never dwell on the past; be grateful for what you have today, work like hell for a better tomorrow, and thank God for the freedom that allows you to do it." He is dedicated to politics and helping to preserve the sovereignty of the United States.

Now retired, Mr. Abbott lives in Goldenrod, Florida, with his second wife, Bernice. He has three children by his first wife, Betty. Jackie, Marsha and Warren have given him six grandchildren and one great-grandchild.

At seventy-eight, Mr. Abbott still enjoys life's hustle and bustle. He is quite the traveler, spending time visiting his real estate holdings throughout three states. He especially enjoys his bed-and-breakfast, The Old Spanish Mission Inn, in

Titusville, Florida, where he maintains the office for the World Liberty Foundation and World Liberty Publishing. The Inn is where he wrote and published this book, and where he is now in the process of writing another. He has a deep passion for writing, be it music or literature.

Mr. Abbott's energies appear effortless and have made him a success to be emulated. His positive outlook on life will continue to inspire emulation long after this book is out of print.